T0195867

Finding Your Way in Science

How to combine character, compassion and productivity in your research career

Second Edition

Lem Moyé

Order this book online at www.trafford.com
or email orders@trafford.com

Most Trafford titles are also available at major online book retailers.

Print information available on the last page.

ISBN: 978-1-6987-1342-7 (sc)
ISBN: 978-1-6987-1344-1 (hc)
ISBN: 978-1-6987-1343-4 (e)

Library of Congress Control Number: 2022920710

Trafford rev. 11/11/2022

www.trafford.com

North America & international
toll-free: 844-688-6899 (USA & Canada)
fax: 812 355 4082

Other books by Lem Moyé

- *Statistical Reasoning in Medicine: The Intuitive P–Value Primer*
- *Difference Equations with Public Health Applications* (with Asha S. Kapadia)
- *Multiple Analyses in Clinical Trials: Fundamentals for Investigators*
- *Finding Your Way in Science: How You Can Combine Character, Compassion, and Productivity in Your Research Career*
- *Probability and Statistical Inference: Applications, Computations, and Solutions* (with Asha S. Kapadia and Wen Chan)
- *Statistical Monitoring of Clinical Trials: Fundamentals for Investigators*
- *Statistical Reasoning in Medicine: The Intuitive P-Value Primer Second Edition*
- *Face to Face with Katrina's Survivors: A First Responder's Tribute*
- *Elementary Bayesian Biostatistics*
- *Saving Grace—A Novel*
- *Weighing the Evidence: Duality, Set, and Measure Theory in Clinical Research*
- *Probability and Measure in Public Health*
- *Catching Cold—Volume 1: Breakthrough*
- *Catching Cold—Volume 2: Redemption*

Lem Moyé

Finding Your Way in Science

*How You Can Combine Character,
Compassion, and Productivity in Your
Research Career*

Second Edition

Lem Moyé
5671 S Wayne
Chandler AZ 85249
USA
LemMoye@PrincipalEvidence.com
https://Principal-evidence.com
Instagram: principalevidence

To Dixie and the DELTS

Preface to The First Edition

If you're a graduate student near the completion of your training, a scientist with both feet firmly set on the bottom rung of your career ladder ready to step up, or a midcareer researcher questioning the meaning of your routine work week, then I want your attention.

As a young man enrolled in college in the early 1970s, I believed that the combination of (1) a good education in science, (2) natural talent, and (3) consistent hard work was all I'd need for professional success. This trifecta would serve as the collision-avoidance system of my career aircraft, flawlessly autopiloting around dangerous objects.

The preprogramming failed as overwork slammed me into one catastrophe after another; and with no sense of purpose and balance, I rammed into the mountains of dissatisfaction, divorce, and disillusionment. While I have overcome much of the damage from these accidents, I am often reminded of the visionless efforts that haunted these early despondent days.

The interminable rain of uninspiring and exhausting activity can drench the junior scientist's early enthusiasm. At first, opportunities abound, and the young researcher plunges in, full speed ahead. Before long, the combined assault of unwavering workloads, exhausting teaching schedules, towering task lists, and miserly pay erodes the researcher's initial zeal. Drained, disoriented, and unable to cope, the scientist begins to withdraw the best of themselves from their work and action that can be professionally and spiritually destructive.

Your experiences during this germinal stage of your career shape your future. For right now, you cannot change the system. However, you can change yourself, moderating the system's destabilizing influence on first you then much later on others who look to you for support and guidance.

Just as today's critical scientific issues are decided by scientists who were once junior workers (afflicted with the concerns and insecurities that you now feel), it is only a matter of time until you yourself play a pivotal role in shaping the impact of science on society. That time is coming for you, like tomorrow's sunrise. When it arrives, you must be ready, meeting it head-on not just with scientific knowledge, but also with strength and wisdom, compassion and vision.

Although character growth is as essential as productivity to the development of the modern scientist, I have keenly felt the absence of a tract written by modern scientists about the role of character and personality in the professional researcher's development. The occasional inevitable missteps that have punctuated my early career had their roots not in some scientific miscalculation but in a neglected character defect. A new sense of value and good judgment, always challenged, but perpetually reaffirmed, along with a set of useful coping skills rejuvenated my career—the acquisition of these traits is the focus of this book.

Lem A. Moyé
University of Texas
School of Public Health
June 2008

Preface to The Second Edition

So here we are, now twelve years after my 2008 edition of *Finding Your Way in Science: How to Combine Character, Compassion, and Productivity in Your Research Career.* What has changed that requires an update.

Two things.

I have, and everything else has.

Since the 2008 edition, I hurdled through a twelve-year NHLBI-funded projects in cell therapy research, becoming team leader of a national coordinating center. That gave me new insight into leadership.

Frankly, I was thrown into circumstances for which I wasn't prepared and, despite my best effort, lack sufficient coping skills. I had to understand how to both learn and engage simultaneously. I and my team made mistakes. I learned how to take responsibility clearly and unhesitatingly. We had both national and international critics whose arguments had to be addressed. Although I learned to build consensus, I also learned lessons that I did not anticipate.

Now, having been retired for four years, I have had time to reflect on my experiences. The Russians have a saying commonly quoted by Stalin: "Whoever looks backwards should have their eyes torn out." Looking back has value only if it can be applied to the future.

The second thing that has changed is everything else. We live in a world with more human beings facing existential, political, zoological, nuclear, and climate threats. Finally, good people are under attack in our time. Public health authorities are spat upon and assaulted. Community leaders with public health mandates are shouted down at meetings. National science leaders are ridiculed. We will put ourselves in the shoes of these people and talk about a strategy that they can employ that combines genuine humility (being completely unafraid of the consequences to themselves) and directness.

So there is much to discuss in this second edition. Think of it as a harrowing roller coaster ride in which Schubert's "Ave Maria" is piped through your headset. It is either scary or comforting. It's up to you.

Lem A. Moyé
Chandler, AZ
November, 2022

Acknowledgments

I have had the opportunity, privilege, and pleasure to work with many junior investigators. Their kinetic questions have energized me, and their intellectual challenges have provided critical illumination as I have walked my own career path. Both their and my trajectories have been altered in the course of our interactions.

I am overdue in acknowledging Jerry Abramson, PhD, and Viola Mae Young, PhD, who have long since retired from the Baltimore Cancer Research Center. These two scientists were the first researchers with whom I worked side by side, and the combination of scientific rigor and compassion I absorbed from them still resonates.

Many have contributed to the content of the book by either inspiring it or by taking time from their schedules to comment upon and criticize its early drafts. Dr. Craig Hanis and Dr. Susan Day provided important advice about the utility of subcontracts for the junior scientist. Heather Lyons, Dr. Sarah Baraniuk, Dr. Wendy Nembhart, and Dr. Sharon Johnatty were especially helpful in their suggestions, convincing me to include important material that I initially did not consider. If the text is instructive and serves you well, then these hard-working young scientists have earned a full measure of the credit. Any problems, misstatements, or inaccuracies in the text's content are mine and mine alone.

These last years, my project managers Shelly Sayre, Rachel Neave, Judy Bettencourt, and Michelle Cohen faced what we believed were towering ethical challenges. These situations excited emotional responses and cerebral ones, and we worked together to not exclude, but integrate the two. Commonly, our own personal beliefs were disparate across our group and we had to integrate those as well.

Finally, my dearest thanks go to Dixie, my wife, on whose personality, character, love, and common sense I have come to rely, and to my daughters, Flora and Bella, whose continued emotional and spiritual growth reveals anew to me each day that through God, all things are possible.

Lem Moyé, MD, PhD
Chandler, AZ, 85249
November, 2022

Contents

Preface to The First Edition xi
Preface to The Second Edition xiii
Acknowledgments xv
Introduction xxv

Chapter 1   Wounded Madness 1

Chapter 2   Trapped 12
             The Ambush 13
             Productivity versus Professionalism 15
                Old versus New Paradigms 15
                Traitors and Heroes 17

Chapter 3   Get Some Rest 20
             Chronic Overwork 20
             Stress-Produced Mistakes 21

Chapter 4   Character Sabbatical 22
             The "Why" of You 22
             Medical School? 23
             The Conductor 24
             Disconnect Reconnect 25

Chapter 5   **Empty Your Mind** 28

    War 28

    An Experiment 28

    Good Servant, Bad Master 29

    Managing the Dogs 29

    Don't Blame Your Mind 30

    Stubborn Thoughts 31

    Congratulations 31

    What to Do with This New Space 32

Chapter 6   **Using Your Sabbatical** 33

    Character and Core Principles 34

    Self-Valuation 34

        Self-Respect 35

    Competence and Survival versus Prosperity 36

        Why Are You in Science, Anyway? 36

        Lives of Others 37

        Inadequacy of the Survival Mentality 38

Chapter 7   **Character Growth** 40

    Source of Esteem 40

    Moral Excellence 45

    Self-Sacrifice 1: General Comments 45

    Self-Sacrifice 2: Sacrificing Time for Another 47

        Gravity 48

    Your Strengths and Weaknesses 49

    Long-Term Vision 51

    Develop Perseverance 52

Chapter 8   **Courage** 56

Chapter 9   **When The Burden Lands** 57

    Individuals in Crisis 57

    Why Does This Matter to You? 61

Chapter 10   **Deploying Courage** 62

    Take Their Place 63

    Why Do Principles Matter? 64

    Costs and Their Anticipation 65

Chapter 11   **General Tools and Skills** 66

    Expand Your Knowledge Base 66

        Managing Zoom and Travel 67

    Air Travel and Family 68

    Working at Home 70

Chapter 12   **Collegiality** 73

    "Educate-able" 73

        Idiot Savants 75

        Be a Colleague to Make a Colleague 76

        Gentle Communicator 77

    Apologies 78

    Manage Your Anger 79

    Conclusions 80

Chapter 13   **Investigator as Collaborator** 82

    Plunge In 83

    Mastery 87

        Learn at Your Level 88

        Systematic Reviews of Manuscripts 89

        The "Dumb Question" 92

        Teach in Addition to Being Taught 95

Reliability 95
   Do What You Promised................................. 96
   Meat Hooks 98
   Above and Beyond 99

Falling Behind 99
   Coping Strategy No. 1 100
   Coping Strategy No. 2 102
   Coping Strategy No. 3............................. 104
   Free Time 106

Accountability.. 107

Becoming vs. Being................................... 109

Engage in the Action.................................110
   Be Persuasive 111
   Sense of Humor 113
   Gain a Sense of the Transient 115

Respect Your Own Judgment.................... 115
   Manage Your Email Content 116

Chapter 14 Administrating Investigator.......................118

The Proposition 118

Administrative Oxygen 119

Experience vs. Expertise 119

Surroundings 120

Institutional Perspectives............................ 120
   Local Environment 121
   Be Familiar with the Rules 123

"Not for Me!".. 124
   Effective Communication 125

Conclusions 130

Chapter 15 Mentorship...131

What is a mentor? 131
   Meeting with Your Mentor 132

Chapter 16   Ethical Investigator 135

    The Scope of This Chapter 135

    Ethics..135

    Ethical Statement.. 137

    Innate Value.. 138

    Undertows 139

    Case Histories..141

        Case 1: Tragic Record.................................. 142

        Case 2: The Patchwork Mouse 143

        Case 3: The 2003 EL61 Controversy 145

        Fraud and Experienced Researchers 147

        "I Promise" 147

    Human Together 149

    Pervasive 150

        Ready, Shoot, Aim 151

    Careful Citations 152

        Deft Thefts 153

        Grant and Manuscript Review Theft............155

        Cite Honestly 156

    Spinning Results 157

    Conclusions 158

    References 159

Chapter 17   Sexual Harassment.................................... 160

    Redline 161

Chapter 18   Ethics and Difficult Superiors 162

    Trust, but Verify 163

    Delegation 164

    Final Comments 165

Chapter 19 Vexing Bosses 166

Castigation 166
 The Intimidation Factor 167
 Listen 168
 Delivering a Message 169
 Personal Abuse 171
 Coping from the Inside Out 172

Chapter 20 Investigator as Leader 176

Seeing Yourself as a Leader 176

Traits of Scientific Leadership 177

Meeting the Challenge 178
 The Twin Challenges to Leadership 178
 The Genius of Leadership 180
 Defeating the Peter Principle 185

Leadership Traits 187
 Knowledge and Resourcefulness 187
 Imagination 188
 Keeping Perspectives Straight 189
 Administrative Skill and Diligence 190
 Boldness and Achievability 191
 Take Advantage of Your Mastery 193
 Developing Other Leadership Traits 195

Source of the Best Ideas 196

The Genesis of Team Leadership 198
 "When in Command . . . Command" 199

Guides for Management 200
 Know the Subject Matter 201
 Know Your People 205
 New Hires 205
 Access 206

Learning to Lose 210

Final Comments on Leadership 212

References 213

Chapter 21 Presenting 214

    Introduction...214

    Sources of Information...................................214

    The Rise and Fall of Stage Fright 215

        The Reasons for Presentation Anxiety 215

        The Response of the Audience 216

        The Hunkpapa Sioux.................................217

    Preparation 218

        The Layman Rule.....................................219

    Walking the Plank 221

    Seventy-two Hours to Go 222

    Reverse Radar 223

    Two Minutes to Go 224

    Giving the Presentation................................ 226

        Countenance.. 226

        Slideshows 227

        When You Make a Mistake......................... 227

        Finishing the Talk 229

        Final Comments 231

    References 231

Chapter 22 Junior Faculty Member 232

    Academics 101 232

    Defining and Evaluating Productivity............. 233

    Teaching 234

    Research.. 235

        Publications 235

        Grants.. 238

    Community Service 239

    Promotion and tenure 240

        Promotion.. 241

        Tenure.. 242

Diligent Days.. 242

    Balance and Discipline... 243

    Developing a Work Lifestyle... 244

Teaching and Character 245

Publishing and Character... 247

Take Charge 250

Chapter 23   **Nova Progenies** 251

Author's Background..255

Introduction

Research is alive. It is exciting. It is inviting. It is hostile and alien. It is intimidating—and sometimes frightening. And it is certainly complex.

In the traditional paradigm, research efforts, while crippled by the absence of technology, were propelled by competent and disciplined thought. Time was readily available for both scientific deliberation and personal development. Today, the twin demands of productivity and communication avidly consume our career time as we take advantage of an ever-expanding world of opportunity. We can perform electronic literature searches using a high-speed internet connection at the airport, followed by an airborne grant-writing session on a notebook computer or tablet while traveling to yet more meetings.

Forty years ago, researchers had to physically stand and wait in line with other researchers to use a computer. Now, computers wait for us, standing ready to word process, calculate, and simulate experiments at our behest. The faster we work, the faster the production of analyses; the more intense the communication, the quicker the generation of papers and product, the more rapid the pace of progress.

In fact, the modern computational systems of science, in concert with contemporary communication advances, can now accept more work product than we can produce. In the old paradigm, travel and technological capacity were the rate-limiting steps. Now we are. The perverse effect of this is that we cannot exhaust the system; we can only exhaust ourselves.

> **The perverse effect of this productivity-communication complex is that we cannot exhaust the system; we can only exhaust ourselves.**

This new dynamic raises questions about the proper role of productivity in science. Unquestionably, a superior scientific product is an important property of a successful scientific career; however, it is not the only important ingredient. The principles, judgment, conduct, ethic, and temperament of researchers must develop along with their productivity if these scientists are to develop into mature professionals.

This is especially true for junior scientists. While possessing fine educational backgrounds, they frequently do not yet have the poise, vision, or coping skills required to both identify and sustain their optimum productivity level.

A philosophical approach that balances them would be a useful foundation. However, junior scientists typically give little thought to developing this set of guiding principles. Unfortunately, many never find their balance and are consumed by or fall away from their chosen field.

Finding Your Way in Science lays out for the scientist how to develop the principles that produce and sustain character growth essential for the scientific professional.

The central thesis of *Finding Your Way in Science* is that the relentless pursuit of productivity ultimately fails the junior scientist.

While productivity is a fundamental attribute of the professional, there are other core themes that must be nurtured and permitted to exert their influence as well. The presence of self-control and patience, of moral excellence and compassion, of discipline and flexibility are as critical to the development of the junior scientist as is the acquisition of technical skills. The presence of these traits engenders collegiality, persuasive strength, responsibility, administrative diligence, influence, and vision, that is, the qualities of charitable leadership.

Chapter 1 provides through a vignette the turbulence and trouble junior sciences can find themselves in.

Chapter 2 articulates the book's foundational theme; productivity, so highly emphasized in academia, in private industry, and in government, is not the only star by which the scientist should steer. Professionalism, which incorporates productivity, is the required cornerstone.

The need to take stock of oneself is emphasized; and the role of self-respect, ethics, sense of charity, and collegiality is reviewed. Concentrating on developing the strengths, skills, and outlook of a mature, professional scientist will not only amplify your productive efforts, but will also buffer and protect you when you face the unseen, inevitable challenges lying ahead. The specific scientific advances you generate will, in all likelihood, be overshadowed and surpassed by the future advances of others. However, the principles for which you stand as both a scientist and as an individual can resonate indefinitely.

Chapter 3 reminds the scientist that rest is essential to the development and growth of their scientific character. You cannot deal with an intricate computation when you are exhausted. Although you may want to or feel driven to work as though you are not tired when you are, this drive is self-destructive.

Chapters 4 through 7 deliver the core instruction of the book where the specific instruction and guidance for character development is provided. A central component is recognizing that "you are not your mind." This recognition opens a new vista through which you can examine and explore your internal morals, attitudes, and sympathies, not just asking what they are, but also why they (and not others) drive you. This leads to regular and frequent character sabbaticals or times that you quietly "inhabit our own character" explore it, challenge it, and play an affirmative role in its growth.

Chapters 8 through 10 discuss one of the main goals and outcomes of these character-building sabbaticals—courage. Building courage requires recognizing that a principled stand that you make can derail or destroy your career and making your peace with that. We rely on our character sabbatical to give us the space we need for this essential growth.

Chapter 11 discusses general tools needed by junior faculty. Concrete guidance is provided on the use of email, and the development of mature scientific judgment is discussed. Approaches that one can use when communicating with and educating senior investigators are developed. Junior investigators are advised to master new scientific knowledge bases and then to take quick advantage of the mastery. However, these strategies work best if they are set of a solid bedrock of self-value that is independent of the approval of others.

Chapter 12 discusses not just the need for collegiality but also how to develop the character trait of self-sacrifice, so essential to professionalism.

Chapter 13 discusses the role of the scientist in collaborative projects. Being the junior member of a research team is a fine opportunity to gain the experience and intuition under the tutelage of senior colleagues. Practical advice is provided for communicating with scientists in other fields whose technical language you do not understand and who may not understand the language of your specialty.

The importance of diligent administration is discussed in chapter 14. This is a topic that many scientists shun because of its nonscientific nature and absence of direct scientific productivity. However, without the development of skill in this area, the scientist runs the risk of inefficiency as she struggles to identify and obtain the critical resources that she needs for her project. Mastering resource management is emphasized. Three focal points are identified that will help the scientist in her first role as principal investigator.

Chapter 15 discusses the importance of developing a solid sustaining relationship with a mentor.

Chapter 16 concentrates on ethical concerns. Examples of ethical failings of scientists in the past are illuminated, allowing the investigator to determine if the seeds of this unethical behavior reside in themselves. Specific corrective steps are recommended to the junior investigator whose superior is abusive or flagrantly unethical. Guidance on recognizing the unethical investigator, how to have discussions with that investigator, and how to foster an environment of high ethical scientific conduct is presented.

Chapters 17 through 19 discuss sexual harassment, unethical leaders, and difficult superiors, emphasizing the importance of character sabbatical in developing the foundation of character courage trait.

Leadership spirit and capability are locked away in the heart of every devoted scientist, and Chapter 20 reveals how the scientist can discover and develop her leadership ability. The importance of (1) taking authority, but not providing too much direction to skilled subordinates, and (2) leading not as an exercise of authority but from a force of honesty and reasonableness are just two of the several strategies for successful team leadership that are provided in this chapter. Junior investigators are exposed to the concept that many scientists who disappoint their own expectations do so because they will not accept, at the controversial moment, responsibility for an immediate decision that they are called upon to make. While investigators can fear rashness, they must also fear irresolution.

Important instruction in Chapter 21 will help put the scientist at ease in presenting results to both small and large audiences. Presentation anxiety ("stage fright") is a real concern for the junior scientist, and its causes, followed by a prescription for its elimination, is offered. The investigator is reminded that the biggest cause of collapse in a presentation is not the intensity of the interchange with the audience but the presenter's own fear of failure. Guidance for preparation and delivery of the presentation is given, and specific suggestions for responding during a postpresentation question-and-answer session are provided.

The discussion in Chapter 22 is specifically for the junior scientist who works in academia. This traditional environment has changed, and the time-tested concepts of academic freedom must now go hand in hand with contemporary concepts of academic accountability. The definition of productivity within the modern academic setting is provided; and the three metrics of teaching, research, and community service are motivated. Since so much of the academician's progress is measured by publications, advice is provided for developing a smooth, positive trajectory for productivity. The concepts of promotion and tenure are discussed. The importance of determining a long-term plan and the need for developing a good and sustaining personal work lifestyle is motivated. As in other fields, productivity alone is insufficient for career development within the university environment. Character development is essential in this setting as well.

Concluding comments are provided in Chapter 23.

Each of the topics in this book is discussed with the goal of not just imparting tactical advice to the investigator, but also to support the general theme that the investigator must develop his or her professional character in parallel with their productivity record. Like the apples of gold in settings of silver, good character and productivity must go together to develop strong scientists.

The audience for this book is broad in scope. It is written at a level for all advanced graduate students, postdoctoral researchers, project managers, and scientists. It is applicable to all scientific fields and to researchers in industry, government, and academic institutions.

1

Wounded Madness

"What's the problem?" the principal investigator yelled into her ear. "You have the file. Do the analysis. Don't get stuck on stupid."

Carol Sinclair, PhD, junior investigator for the Hendricks team, jerked the phone from her ear as the car bounced over the rough roads. *I've been a devoted team member for years*, she thought as the principal investigator of the project continued to pour invective through the phone, tying her stomach into acid knots.

The twenty-nine-year-old closed her eyes, her right hand gripping the Uber's torn armrest as the car bounced over the rough airport roads to DFW and the plane that she didn't dare miss.

The scientist inhaled then brought the phone back up. "I'm doing all that I can, Dr. Hendricks. I'm on the way to the airport now. Please understand that there is just no time. I"—she stopped in midsentence, a sharp bump of the taxi suspending her in midair for a nauseating moment—"can't complete the a—"

"Stop whining and do what I ask!" he shouted. "I don't care how or where you get it done. Just finish the work."

Click.

Carol leaned her head back against the Uber seat, blocking her eyes from the bright glare of the Dallas morning sun that burned through its dirty windows. She dropped her cell phone onto the seat next to her, moaning while raising her pale slender hands up to stroke the sandy-brown hair covering her temples.

But the headache would not be rubbed out.

She shook her head then checked her watch: 10:00 AM.

Plenty of time to catch the twelve oh five. She'd be as good as dead if she missed it.

●

Earlier, the alarm blared through the darkness. She'd staggered out of bed at 4:00 AM after four hours of sleep, navigating her way through the drafty apartment with its cold floor to the old notebook computer perched on the cheap desk.

Postponing the bliss of a hot shower, she plopped down onto the seat, the cold of its cheap metal startling her. Shivering, she stabbed a key to awaken the hibernating machine.

There she had sat in the glow of the fifteen-inch display, reading each of the complex messages. Zoom call schedules. Budget requests. Concerns about lost lab specimens. University questionnaires demanding updates on her quarterly productivity.

Each was critical, screaming for a reply. She sped-read the messages, keying in responses in rapid-fire succession. Updating data spreadsheets. Copyediting abstracts. But Carol could only focus for a short time this early morning. Just too little sleep. Eyes burning. Messages blurring.

She jumped at the sudden rumble of a passing truck then checked her watch.

Seven fifteen.

Damn.

The five-foot-seven-inch scientist closed her email client then rushed to the bathroom where she roughly washed herself during a tepid two-minute shower.

On the way to her closet, Carol paused for a moment to look in on her sleeping baby. Gazing at the one-year-old for whom she had little time, a wave of despair rolled over her as she rushed back to the bathroom to comb her hair. The assistant professor frowned at the new worry line etched into her forehead.

By eight thirty, she had pulled on a gray sweater and black slacks. Slipping into a pair of flats, she hurried into the kitchen.

"Another trip today?" Jacob called out from behind his newspaper. "Third one in two weeks, right?"

He'd been counting.

Carol's stomach clenched, and she struggled against the wave of panic that roiled up.

The scientist walked back from the coffeemaker, forcing herself to breathe slowly, her mind's eye scanning for a reply that would keep the morning on an even keel.

"Jacob," she began, gripping the steaming cup of coffee as she sat next to him at the small table. "I think I told you about this presentation. Our group's results are exciting, and the pre—"

"Oh please, Carol," Jacob snapped, jerking the newspaper down to reveal dark eyes whose ferocity startled her. "Spare me. This is the fourth 'pivotal' presentation you've had in two months."

"Lots of travel, I know, but success comes in spur—"

Her hand jerked, spilling coffee onto her sweater at the sudden blare of the waiting Uber. "We'll work this through when I get back," she said, hurrying over to get some water for the spreading stain.

"What's the point." He snapped a page of his paper. "You are either preparing to travel, traveling, or recovering from travel. Those are your three states now, and your family's not in one of them," Jacob said, voice rising, "Is there any time for good conversation about you, us, or our baby?"

"I can't help this now, honey," Carol said. She stepped across the small kitchen toward him, hoping for a goodbye kiss. "I really need to catch this flight."

"Really?" he asked, pushing back from her. "Let's put this to the test. This time, say 'No'. Just don't go."

"What? Jacob, they rely on me."

She jumped as her husband slammed the paper on the small kitchen table. "Rely on you? Or use you?"

He turned in his seat to face her, almost falling off the breakfast chair. "You know, Carol, the world will continue to exist in all its complexity if you were to get sick today."

The tears came from nowhere. "If I don't go, I'll get fired."

"No, Carol, you won't. You just think you'll be fired." He stood. "How many nights have you come home late? How many weekends have you spent in the lab? You've earned the time off."

"That's just the way it is."

"It's that way because you tolerate it. Listen." He sighed. "I get that you have to work hard sometimes. I do not," he said, moving his hand, knocking over the milk carton, "accept that you must do it all the time."

The horn blared again.

"I have to—"

"No, you don't," he said. He strode to the front door, yanking it open to see a middle-aged man halfway up the steps to the apartment.

"You all want Uber or not?"

"Take your car and get lost."

Carol watched the two men stare at each other. Jacob, balling his fists, took a step down toward the driver.

"Please," she called from behind her husband, "we'll call another."

"Yeah?" The man backed down two steps, looking at Jacob, then turned. "Good luck."

Carol watched her husband stare at her. She saw that his face was flushed, his eyes narrowed.

"I'll fight for you, Carol. But you don't fight for us."

"Jacob, I have to go." She pulled her cell out and made another Uber contact.

The scientist saw Jacob gaping at her, eyes and mouth wide open. Then he pursed his lips.

"Get the hell out of here," the husband said.

Carol's heart skipped a beat at the insult, and she froze in place for a moment, her eyes filling with tears. Jacob blinked twice and leaned closer to her, his mouth open again, eyes softer.

"I—"

They both jumped at the horn blast. Carol pecked him on the cheek then walked to the door. "I'll phone tonight. We can talk then. I won't forget to call the sitter about the keys."

Jacob shrugged his shoulders and walked back from the door to the kitchen.

Carol scooped up her overnight bag, notebook computer, and purse. Running outside, she jumped into the dusty backseat of the old vehicle, resigned to yet one more slow commute to DFW this month.

●

The taxi suddenly changed lanes and screeched to a halt at the airport departure entry. Jarred back to reality, Carol extended her arm out in front of her to buffer the jolt. Snatching her belongings from the seat, she bolted from the backseat of the cab, hurrying across the dirty pavement into the terminal.

Gotta go. The faster she collected her boarding pass and got through security, the more time she had to complete that last data analysis before her plane took off.

Fifteen minutes later, Carol shifted anxiously from one step to another, waiting in a meandering line of passengers waiting to talk to the ticket agent. Each customer took forever, explaining their confusing circumstances in halting language.

Blinking rapidly, the scientist struggled to control her seething anger. At this rate, she'd never get to the gate in time to get some pressing work done.

Finally.

The head of the line.

"Listen," she began. "I really want to be as patient with you people as pos—"

Her cell chirped. She shot it a glance.

Hendricks.

Her stomach jumped as she declined the call while focusing on the ticket agent. "The kiosk wouldn't let me check in automatically. I just wasted twenty min—"

"Just a moment," the agent replied. Carol watched the frowning employee study her hidden terminal.

"You can't check in automatically if you're within thirty minutes of departure, Ms. Sinclair." The agent looked up. "Your flight leaves in twenty minutes."

Carol froze, staring in open-mouthed astonishment. "What are you talking about! It's only 10:45 AM. My flight's not until 12:05 PM!"

"Actually," the flight attendant said, shaking her head, "your flight is scheduled for departure at 11:05 AM." She clicked across more keys.

The scientist heaved, as she realized that she had misread the flight time, the second time in two months. The huge open airline terminal spun around her while an icy dread closed around her heart.

"Ms. Sinclair?"

"Yes," Carol answered, shaking her head, refocusing on the agent.

"The next flight is 1:15 PM. There's one seat left if you want it."

"Uh . . . sure . . . thanks. That'd be great." She took her ticket and walked to security.

She knew that the delay meant she'd miss the rehearsal with Dr. Hendricks and the research team scheduled for later that afternoon.

But extra time in the terminal would let her get some work done. She knew she'd miss the rehearsal, but at least the last-minute analyses he requested would be complete. She exhaled slowly and closed her eyes.

Twenty minutes later, Carol plopped into a seat at her gate. Ignoring her hunger pangs, she fired up the laptop perched on her touching knees. She counted the seconds as the machine booted up then searched for and found the airport's wireless signal. A moment later, Carol anxiously scrolled through her new email messages.

There was another message from Hendricks. She held her breath as she opened it, rapidly scanning the PI's terse note for signs of a new crisis.

No new emergencies. Relieved, she dashed off a message to him explaining that she was coming in on a later flight, then arranged her Uber car for Denver.

Relaxing some, she clicked on the second of twenty unopened emails.

The message refused to open.

Carol clicked again. The client opened in slow motion, the hard drive wildly clattering. One second later, a single message exploded onto the screen.

Trojan agent found in C:\Documents and Settings\csinclair.

The computer viral infection meant she wouldn't complete her work. The young researcher groaned, slumping back into the uncomfortable seat, her arms dropping by her sides. She turned and noticed a passenger, two seats away, gaping at her.

Carol chewed her lower lip as she took a few deep breaths, fighting the stomach spasms back. Tapping the keys carefully, she coaxed her crippled machine out onto the internet, where she found and slowly downloaded a new virus-removal tool.

Three hard drive scans and two reboots later, her machine was back to its normal speed. The scientist sighed as she drilled down the subdirectory tree to find the data file from the last experiment.

"This is your preboard announcement," the gate agent called over the static-cluttered speaker just above her seat.

Carol jumped at the grating voice.

Her impromptu work session was over.

Nothing to show for it.

The junior investigator joined the long line of passengers making slow progress down the narrow Jetway, her exasperation and fear merging into Superstorm Headache. She made her way back to row 33, hoping they would have some Tylenol on the pla—

Babysitter.

She hadn't changed the arrangement with the babysitter, who would need to know the new key location.

The new pain band was almost audible as it snapped tight around her head.

Carol fumbled her luggage from one hand to the other, as she fought to get a glimpse of her watch.

One twenty-three.

The babysitter would arrive in less than ten minutes.

Pain band tighter now.

Panting, the scientist pushed her way back to her aisle. Cramming her luggage in the overheard, she dropped into the narrow middle seat, yanking the phone from her purse then panic-dialing the babysitter's cell.

Busy.

Her eyes rolled, as she whimpered to herself. Dialed again. Busy! Jacob will be furious, at the lapse. She dialed and re-dialed as if she would force her sitter's phone to accept the call, overriding all else.

Just text he—

"Hello?"

"Nancy? Nancy! Thank God! This is Carol. Listen . . ." Carol conveyed the new instructions into the phone, shouting over the flight attendant's loud announcement that it was time to turn cells off. Carol finished the message just as the heavy cabin doors thudded shut.

Once in the air, she popped open her laptop. Holding her breath, she clicked on her work directory. It opened smoothly. No sign of the virus.

Subsisting on diet soda and peanuts, Carol worked furiously on her PI's last-minute analysis and then its summary. She stared hard at the document on her screen, working and reworking the sentences.

"Ma'am."

"Yes?" Carol answered, jumping at the interruption.

"We need you to turn your computer off. We're in our final approach now."

"Sorry. OK."

She powered the computer down, shaking her head in frustration. Not there yet, but almost done.

As soon as the plane's wheels skidded onto the runway, she popped her cell on.

Three missed calls.

Hendricks. Hendricks. Hendricks.

Fighting the impulse to hurl the phone to the ground, she trapped the cell between her neck and ear as she schlepped her luggage to the jetway.

"Damn it, Carol!" he shouted. "We started our rehearsal two hours ago. Where the devil are you? And where's that analysis we need?"

"Didn't you get my email, Doctor? I had to catch—"

"I got nothing. For heaven's sake, Carol. You know how important tomorrow's presentation is. Do you have the analyses or don't you?"

She could almost feel his spit flying into her eyes and mouth.

"Uh, not yet . . . I'm almost—"

"I told you we needed that."

"You just told me this morn—"

"So you can't find an hour to answer these critical issues? Is that what you're telling me and your research team? Don't embarrass us with your incompetence."

Carol began to shake. "I . . . I worked on it on the plane. It's almost—"

"I have no time for you now!" her boss shouted. "It's bad enough that you've missed most of the rehearsal. The least you can do is have the analysis done when you get here." The phone banged in her ear as he hung up.

The insults, delivered with the pain of a rock jammed up her nose, blurred her vision. Hot tears fell from her cheeks. Carol leaned against the cold, folding jetway wall, breathing rapidly as the rough passengers banged into her, pushing by.

Ramming the phone back into her purse, she raced to the Uber line, falling into the backseat of the dirty car. Gruffly pushing aside the desperate need for sleep, Carol opened her laptop, using its last few minutes of power to push her way through the analysis that her boss demanded.

●

Rain delayed the Uber's arrival at the cheap hotel. Carol stepped out, being sure that the laptop, purse, and carry-on were in hand, and rushed through the lobby check-in. She tempted herself with a bath, but would have been happy to settle for a three-minute freshening u—

"Carol?"

She whirled to see Dr. Fred Filar, one of the coinvestigators in the study. "Oh, Dr. Filar! Hi. I . . . I just got in, and I'm—"

"Dr. Hendricks sent me over to meet you as soon as you arrived," the senior investigator interrupted, hands folded over his chest. "You're late." His angry stare leaped out from the narrow eyes just above his hawk nose.

"I know," she replied. She ran her hand through her hair, fidgeting for a moment. "I need to go upstairs for just a second, I'm—"

"Come on, Carol!" Filar said. She watched him drop his hands to his sides, balling and unballing them. "There's just no time for that now. Our rehearsal isn't over yet, and we have lots of questions for you." He grabbed her arm, pinching the skin above her right elbow, turning her around.

"We're going to our 'war room' across the street. Plus," he said, reaching into his brown jacket pocket, pulling out a minidrive, "you'll need this."

Carol's face twisted, and she breathed rapidly as the frustration swelled up like a balloon. "Dr. Filar, it should only take me a few minutes to check in and drop my things off in my room. Please. Can't I do that first? I promise that I'll be right—"

"You are already late. Let's go." Filar started walking; and Carol, head down, hurried to keep up with the senior investigator as he strode out of the hotel. "And take this drive with the next paper version."

She took it from him as they walked outside then dodged cars in the wide avenue when the light turned yellow.

"We decided to submit the results manuscript while we're here at the meeting," he called out when they made it to the curb. "Can you do that for us tonight?" They walked into the Semos Hotel just as a pouring rain began then down a dark corridor to the left. "It should take you just a few minutes electronically."

She stopped, her pulse racing. "What? Are you telling me that paper has to be completed tonight? Including the new analysis? That's impossible."

He turned to her, scowling. "Why?"

"First, the analysis on a laptop will in and of itself take an hour. Then the analysis has to be interpreted by you all, and the manuscript discussion section rewritten. It took two hours for the team to agree to the last discussion section change. And finally, the electronic submission itself takes hours, not minutes. Confidentiality statements. Signatures from all coauthors attesting to their roles in the work. Financial disclosure statements." She swallowed hard. "It won't happen tonight."

Filar held open the door for her into the conference room where the research team was meeting. She heard Hendricks's voice.

"Don't tell me, Carol," Filar said, a sneer emerging between his cracked lips. "Tell him."

●

Four forty-seven AM.

Carol turned away from the computer screen, eyes burned by scratchy fatigue fire.

Wearing only bra and panties, she stood up, letting her eyes roam over the mess of the drab hotel room. Soggy clothes lay in a dirty heap at the entranceway. Research papers were strung everywhere. The room reeked with the smell of perfume leaking through her unpacked overnight.

Suddenly, her headache jumped at her, large gray spots appearing before her eyes, the taste of metal filling her mouth. The scientist fell forward onto the bed and, head hanging over the edge, reached for the small circular wastebasket that was just within her grasp.

Knocking it over, she vomited up and down her outstretched arm.

Carol lay there for a nauseous moment while her stomach's painful heaving gradually subsided.

An unmitigated disaster.

Team meeting lasted four additional hours, filling the evening with hour after unending hour of more "emergency" analyses. She didn't check into her hotel room until after ten.

Statistical evaluations consumed two hours. Rewriting the discussion section (and identifying the need for an additional "last" analysis) took another two hours. The hotel's slow and quirky internet connection consumed a full two hours to upload the manuscript and all the requisite forms and acknowledgment and signature pages.

Jacob's vituperative call, demanding why their cell phone bills hadn't been paid last month, capped the calamitou—

Ding.

She looked at her phone.

Hendricks.

The fifth message from him since the meeting ended. She watched the blinking icon's ceaseless demand for her attention through wet eyes, ignoring it.

After several moments, Carol got up from the bed and staggered to the bathroom to wash her stinging arm. Toweling it off, she walked to, then past the small desk where the computer sat, across to the stiff upholstered chair where an unopened pack of cigarettes lay.

The phone diode blinked away as Carol opened the pack, extracting one of the long, slender cylinders. She walked to the window, opened it wide, then lit one up, taking a slow luxurious, drag, the smoke deep in her lungs.

After a moment, she deliciously exhaled, dropping the rest of the cigarette from the window, watching its buffeting by the wind as it drifted to the cold cement fifty feet below.

She turned and stared at the computer screen across the room, as she relished the nicotine remnant in her lungs. Then she walked over, unplugged it, and holding the whirling machine by the dirty screen till she got to the open window, flung it out as well.

The only career she'd ever wanted was one in science. She'd ached for it, prayed for it, driving herself hard toward it.

Shaking her head, Carol turned her back on the window as she reached into the dark room, pulling the decision that she needed to make for weeks now fully into mind.

The investigator transmitted a two-line Hendricks message on her phone. As she typed, a remnant of an old favorite quote of hers jumped onto the racetrack of her mind, running round and round.

"The heart knows what the heart knows . . . The heart knows what the heart knows . . . The heart . . ."

Carol jabbed the Send key then walked back to the window. There, as she watched the driving rain pelt the city, her tears flowed freely, following the smooth contour of her cheek to her chin, where they dripped onto the black windowsill.

Her cell phone chirped once then repeated itself in three minutes, then again a minute later.

Hendricks.

No more. She found the website of Southwest and arranged a flight for noon.

Time to flee this terrible, wounded madness.

Time to go home.

2

Trapped

Let's be clear. No rational person plans for that cutting, disruptive experience just described. In fact, it's amazing, tragic, and alarming how many young scientists choose to inflict it on themselves.

Some say that this is what to expect in order to make progress in this competitive field. Others say it's just a short-term sacrifice. Some claim they actually enjoy it.

Yet few young scientists are actually prepared for it.

They understand that science requires both careful planning and frenetic activity, deliberative thought and incisive action. It rewards the quick, audacious choice of the right solution and rapidly leaves those who cannot keep up behind.

They get this.

However, what they don't see is that this labor is often generated in a cauldron of tumult and emotional upheaval, a bruising delirium that can disorient, confuse, and dishearten you.

As you start your career, you may be buoyed by a natural optimism. With no real understanding of the challenges ahead, you're filled with enthusiasm and excitement. After all, didn't your combination of intellect, computational skill, and promise guide and sustain you through graduate school?

And wasn't one of the points of graduate school to deliver you here, to this moment, for this new job? Certainly, those natural abilities that guided your path thus far will not fail you, right? Past successes should produce future victories, shouldn't they?

For others of us, the new job's confirmation brings foreboding. You know that you'll do your best, but fear of the unknown overwhelms your fledgling confidence. Just like the anxious bull rider who holds tight to the grip on the angry animal's neck, you dread that once the gate swings open and the ride starts, the early calm will be broken by a few sharp, quick bucks followed by a hard fall.

Ready or not, optimistic or not, you're about to be overwhelmed. You need some new skills to survive and prosper on this new ground, and you need them fast.

The Ambush

Whether you're a statistician working in a clinical trial, an epidemiologist in a government surveillance group, a computational biologist working on reverse engineering an antibody's structure, a behaviorist engaging with the community on a research project, or a new assistant professor of physics at a university, the dizzying pace of activity is overwhelming.

New scientists are in great demand. While you are commonly expected to develop your own scientific agenda, your skills will be in demand to support additional projects.

The principal investigators on these other projects, with their own demanding production schedules to meet, are eager for you to pitch into the research fray. They require your quantitative competence, clarity of thought, and communication skills. Quickly, and through no great effort of yours, you'll find yourself on two or three projects within the first few short weeks of your new job, rapidly immersed in the grant applications, interim reports, abstracts, and manuscripts that reflect the effort's productivity schedules.

Instructional obligations will also be yours to bear if you're part of a teaching center. The current shortage of teaching scientists means heavy teaching loads for faculty. If you're lucky, there will be a short probationary period during which you'll be permitted to strengthen your didactic skills under the aegis of a senior instructor. However, whether you're lucky or not, you'll soon be responsible for teaching at least one course on your own.

Specifically, you'll be standing or Zooming before your students, giving lectures that you yourself have planned. You will also be offering tutorial sessions, grading assignments, evaluating examinations, and determining final grades. In addition, you'll have to provide time (and wisdom) for your students.

And on top of all this, you must navigate the treacherous waters of institutional or university politics as you serve on different boards, councils, or committees. These groups commonly have their own Twitter, Instagram, or Facebook feeds, in addition to the voluminous university email.

You master new technology. Complex computing tools are now within the budgets of most scientists, and its accessibility means that we work in an environment of ubiquitous calculation—we can compute anywhere. In fact, if you're a quantitative scientist with your career affirmatively based in computing and electronics, your colleagues will expect you to have the greatest adeptness with these new tools since they are, after all, "part of your specialty."*

Ever-present connectivity goes hand in hand with omnipresent computing ability. You can reach out (and be reached) electronically; instant text messaging makes you instantly accessible.

The use of attachment-incorporating email permits you to interchange sections of a grant with colleagues around the nation (and the world) from your bedroom at home, the beach, or your hotel room on the road. Students can reach you via email, class chat rooms, and electronic blackboards; your coinvestigators can hold Zoom conferences on your smartphone. You can develop and refine a slide presentation continuously, right up until you stand up, Wi-Fi connect your smartphone to the projector, and actually give the lecture.

You can do it all, and do it anywhere, all the time.

That's the great trap.

The combination of ubiquitous computation and instant communication underlies a common theme of the twenty-first century. You, as a scientist, are blessed with (1) a stupendous wealth of productive opportunities and (2) new and modern tools to support your pursuit of these opportunities. There are always new projects to take part in and new ideas to generate. Your marching orders are to wield these tools in full-fledged pursuit of your research agenda.

You are to produce, present, proselytize, publish, and promulgate.

The ambush is that you can never generate enough productivity that will satisfy a system that absorbs more than you can ever hope to

---

* This discussion sets aside the additional consistently required efforts on your part to keep up with software upgrades, hardware maintenance, and securing your work from viruses, worms, Trojan horses, rootkits, spyware, and "pop-ups" that plague computer use and the web.

produce. No matter how productive you are, you will miss opportunities for even more productivity.

You cannot slake the productivity-communication thirst of the research system.

> **Thus, the pleasure of being able to work anytime devolves to the drudgery of working all the time, and the opportunity to work anywhere reduces life to an existence of working everywhere.**

Productivity versus Professionalism

A television commercial that was popular sixty years ago used to ask its smoking viewers if they were "smoking more, but enjoying it less."*

Scientists in the twenty-first century might be disturbed by their own honest answer to the relevant question, "Am I working more but enjoying it less?"

We are more productive, but productivity for its own sake can produce chronic dissatisfaction. Are productivity and the mastery of technical skill all there is to being a scientist?

Old versus New Paradigms

While the evolution and expanded availability of technology have been undeniably beneficial, our wholesale embrace of it requires us to give up something of value. This subtle exchange is almost immeasurable on a daily basis. We see the consequences of this exchange over months, not hours.

As a college student in the early 1970s, rapid and easy access to high-speed computing was impossible.** We shared computer time on a single large mainframe machine that served the needs of all students, faculty, and administration for the entire campus. Since only one or two of my quantitative analyses could be carried out each day, each of these analyses

---

* http://tobacco.stanford.edu/tobacco_main/images_body.php?token1=fm_ img5012.php.

** Computer time was allotted for the entire class as a whole—not for individuals. For example, each class section (composed of approximately twenty students), would receive thirty minutes of computing time for the whole class over the complete semester. If the computer time was "hogged" by a small number of students, thereby consuming the class's time before the semester's end, the class could not complete the curriculum. In order to avoid this, the class's computer-time consumption was closely monitored by the professor.

had to be planned carefully. I was forced to check and double-check my programs since computer time couldn't be wasted. However, because computer sessions were spaced far apart, it was easy and natural to take the intervening time to think carefully and critically.

In addition, I commonly had to physically stand and wait in line for a half hour with students and senior scientists to use the single computer. No smartphones, no tablets, no texting. We as colleagues were electronically detached, so we connected to one another. This provided even more downtime, affording me the opportunity to think about and sometimes discuss my problem with others, be it a chemistry professor or a freshman. The conversations were leisurely, relaxed, and instructive.

There were also other restrictions that cleared the way for careful, deliberate thought. The high cost of flying combined with the relative dearth of meetings constrained scientific travel. In addition, since the number of journals in any particular field remained small, it would have been foolhardy to try to publish a peer-reviewed manuscript each month.

The torpor and inefficiency of this old paradigm is clear. However, it had the strengths of its weaknesses. The built-in "downtime" provided ample opportunity to think carefully about problems. There was time for daily discussion and reflection with others who were both junior and senior to me. This time was of great value. Critical thinking was valued, with time being available for it.

Unfortunately, by sweeping away these barriers to efficiency, the technologic breakthroughs in computing and communication have overrun the built-in time for reflection and assessment. This contemplative time is still of great value, but it no longer comes naturally. It's no longer waiting at our feet. We have to actively seek it.

> **Modern science and invention have removed many technological obstacles. Unfortunately, by sweeping away these barriers to efficiency, they have also overrun the built-in time for reflection and assessment. Contemplative time is still of great value, but it no longer comes naturally. We now must actively seek it.**

Yet the need and availability for reflection are difficult to see in the new scientific technology-dissemination system that can now accept more work product than we can produce. The faster we work, the more experiments we conduct, the faster the production of analyses, the quicker the development of papers and products, the more rapid the pace of progress.

And despite the Herculean productivity effort on our part, we do not saturate the system. If you were to work twenty-four hours a day, seven days a week, week after week, month after month, sacrificing both you and your family in the process, you would still miss opportunities to read, theorize, compute, experiment, travel, and write.

We cannot exhaust the system. We can simply exhaust ourselves.

And what of restorative quiet time? It is rejected by the modern product-promulgation science system. Peaceful reflection, at first glance, does not fit with the modern definition of productivity. It does not consume bandwidth. You cannot add it to your curriculum vitae. It does not increase your Twitter or Instagram followings. It does not produce new submissions to new journals or novel presentations in never-before-heard-of but available conferences.

An afternoon of quiet thought, seen from the constraining perspective of a modern technology-dissemination system, adds no value.

This is the source of that gnawing restlessness that we feel when we are not active. We have come to equate a quiet, "still" time with missing an important opportunity to be "busy," to be productive. This deceptive philosophy is destructive since there are more opportunities than you can ever hope to seize. And if you do not seize control of your role in the system, then it will certainly seize you.

Traitors and Heroes

Productivity is what you generate. Character is who you are. Unfettered productivity in the absence of solid character is woefully inadequate, as demonstrated by the contrast between two contemporaneous figures in US history: Benedict Arnold and George Washington. Admittedly, neither of these men has any scientific record of note.

George Washington is revered for his role in the birth of the United States. Cities, streets, and monuments commemorate him; and his face adorns US currency.

However, no city or state is named for Benedict Arnold, and Arnold has no monuments in his name—only a small and unremarkable grave in England. Not only was he guilty of treason, but Arnold is also the archetype US traitor, betraying his young country when she was in desperate need of capable and loyal officers.

Yet the calamitous fury of the Revolutionary War in its early years predicted a different outcome. General Arnold performed spectacularly for the colonies. In fact, for a time, Benedict Arnold was the only winning general the colonial army had. He won tremendous victories in upper New England, captured Fort Ticonderoga, and almost seized Quebec for the colonies. In addition, Arnold organized and supervised a freshwater navy that served well on the critical New York lakes. These were remarkable achievements for an army officer.

Arnold's accomplishments stand in stark contrast to those of George Washington, whose campaigns were mired in mismanagement and defeat. After Washington withdrew from Boston, he transferred his army to the middle Atlantic states. There, his strategic blunder led to the route of his army on Long Island by a relatively timid British general.

Moving west, Washington lost in rapid succession Brooklyn, Manhattan, and Harlem, ceding New York City to the British for the remainder of the war. After two astonishing victories at Trenton and Princeton in 1776–77 (that surprised even him), Washington quickly lost several additional battles in New Jersey. Transferring his shredded army to Pennsylvania, he promptly lost Philadelphia to the British for the war's duration.

This string of defeats was unprecedented for a commander in chief, generating lacerating criticism of the American leader.

Meanwhile, Benedict Arnold was held up as a fine example of an American officer. Possessing a good strategic sense, he could keep the tactics of a fluid battle clearly in his mind and was personally courageous on the battlefield. It was Arnold's two victories at Saratoga, New York (during which he was seriously wounded), that finally convinced the French to enter the war on the side of the colonies, a decision that slowly turned the tide of the war in favor of the Americans.*

Yet Benedict Arnold, productive general that he was, became the quintessential traitor while George Washington, the hapless and nonproductive commander in chief, became and remains a national hero. How did this happen?

While there were undeniable differences in productivity between these two important military leaders, it was their stronger differences

---

* During this frenetic period of negotiation, the British tried to end the war by offering favorable terms to the colonies, but news of the French treaty of support quickly undermined any positive American response to the conciliatory British offers.

in character that emerged as they faced the two inevitable reactions to people such as them who wield power: (1) undeserved adulation and (2) unearned, unremitting criticism. Each reacted according to his personality and values.

Benedict Arnold, despite his battlefield accomplishments, was unable to buffer himself from severe criticism for his inevitable small errors. He reacted harshly whenever he was denied a privilege or passed over for promotion. His anger at these denials and slights hardened into intense resentment first and then hatred of those whose cause he championed so well. Losing himself in this agitated state, the general fell into the traitorous scheming for which he is solely remembered.

Alternatively, George Washington was grounded by a character that buffered him from the warping effect of his ghastly record's painful criticism. The fact that the country and the army stood by Washington is a tribute to how they appreciated his character. As Morrison stated, "In no other revolution has a loser of so many battles been supported to the point where he could win."[1]. Rising to become president, his character continued to grow, now effectively buffering him from hysterical adoration.

Therefore, while a difference between these two men was productivity, the difference that mattered was the difference in character. One was able to grow. The other was not.

Productivity is unquestionably important. One cannot have career development and advancement without a solid productivity track record, a track record that requires consistent, patient, hard work. However, although productivity is necessary, it is not sufficient.

While work yield is a high priority for you, your character development should take precedence. You can encourage this by developing a balanced plan for professional growth and maturity. This plan must incorporate productivity, but also emphasize other contributors to your career growth. The next chapter identifies these core areas.

---

1   Morrison S.E. (1972). *The Oxford History of the American People. Volume 1.* London. New American Library, p 328.

3

Get Some Rest

Chronic Overwork

Before you begin a "character sabbatical," you must be rested. The sabbatical will do you little good if you spend much of it sleeping.

Since I probably don't know you personally, I can't say how much of your work lifestyle will be adjusted if you follow this book's advice and monitories. However, there is one thing I know that you need. Regardless of your education, knowledge, training, experience, and expertise, you must be well rested to bring your talents to bear on the problems at hand.

One of the most perniciously destructive actions that will ruin your experience as a scientist is chronic overwork. Just as acid will burn through your clothing, overwork eats through and ultimately shreds your ability to do anything effectively. Fatigue breaks the link between you and your unique combination of talents and abilities. Important memories hover just out of your reach. Weariness makes you short-tempered and sick. Regardless of the degree of control that you have over your own nervous faculties, you must rest.

We all readily acknowledge the harm produced by overwork and insist that our loved ones be rested. Perhaps the resilience of youth protects you from your intensive, sometimes brutal production schedule that denies your body its needed restoration. However, even among the strongest, the need for rest becomes acute.

Actively build sleep and rest into your schedule. Just as you insist that there be time in your schedule to teach your class or to go to important management meetings, be equally diligent about insisting on getting the rest that you need. Invest the same energy into taking care of yourself as you do with the development of your product or the design of your experiment. Look at your calendar, not with the view to crowd out these rest periods, but to protect them.

Stress-Produced Mistakes

In fact, during times of stress, you'll need to lengthen them.

The unpredictable changes in your calendar may require a flexible approach to sleep. If naps work for you, take advantage of them. If you need to find someplace quiet, then seek it out. If you have a challenging three days before you, insist on getting more sleep both before and after the stressful period. Whatever your pattern of sleep is, do not neglect it.

Additionally, fight for the time that it takes to exercise and to eat properly.

Of course, an obvious argument against the thesis that we need sleep is that it takes time that we don't have. However, this contention ignores the fact that the productivity gained by sleep-deprived work effort is commonly overturned by a collection of mistakes that are more likely to occur when we work without rest. Incorrect reasoning more easily penetrates the fatigued mind and produces a thought process that while attractive when we are tired, reveals its foolhardiness in the bright light of a day.

The world will continue to go on if you get another hour's worth of sleep.

"Productivity" generated by overworking is commonly overturned by a collection of mistakes that more easily occur when you work without rest.

The more stressful your job, the greater the requirement for rest and disconnection. All too many of us learn this the hard way. Don't be one of them.

4

Character Sabbatical

The "Why" of You

We have all heard (and may be envious) of a senior faculty member who is "on sabbatical." Free of the worries of daily academic and professional life, they are at liberty to pursue an intellectual interest of theirs.

Sabbaticals typically last for an academic year. The activities can be unrelated to the faculty member's grant-supported research topic, and the scientists do not have to attempt to dovetail what they have newly learned with the focus of their prior research activities.

While commonly fulfilling, these sabbaticals rarely produce a tectonic shift in the core belief of the senior faculty member. The sabbaticals that we are talking about in this chapter are shorter, powerful, and enabling, reaching to your essence.

A character sabbatical is a block of time that you take by yourself to plumb the depths of who you are. Its specific goal is to bring you to turn over your fears, beliefs, motivations in order for you to see, to touch, and to rebecome the "why" of you.

You will take hundreds of these sabbaticals through your life; and they will adjust, alter, and steady your career.

An example of what happens without this focused self-evaluation is the example of Dr. Sinclair in chapter 1. Her attention, so focused on her work, marginalized the rest of her life, turning her essential nonwork necessities into bothersome interruptions of her job productivity.

In addition, because she never knew to take the time to separate herself from work, Dr. Sinclair missed the opportunity to see what was happening to her.

Retaining focus on the "why of you" is essential because life is a rending experience. It is not linear, but chaotic with sudden changes in trajectory. There is no purpose in staying focused on a path when the destination has changed. Without paying attention, you miss the new goal.

Medical School?

Consider a twelve-year-old student focused on the idea of going to medical school. Announcing this in middle school, she sedulously worked into and through high school and college, her mind clear on her goal, but never questioning it. She had no family member in medicine and had only fleeting conversations with her physician when she saw him once or twice a year. She did no research on how physicians spend their time, assuming that if one was a physician, one must automatically enjoy being a physician.

As a senior in college, while completing the voluminous medical school applications, she reviewed her stellar transcript. Yet on deeper reflection, a semester-by-semester evaluation revealed that she truly enjoyed her mathematics courses more than anything. In fact, she disliked her biology premed curriculum, suffering through their teachings year after year.

But she did not reexamine her goal, seeing her struggle in biology courses as the "the necessary sacrifices to achieve the goal." Finally, she was admitted to medical school.

She abhorred it.

Astonished at her reactions but determined that they were wrong, she struggled through the courses. While the copious amounts of data and material was overwhelming, she was not really challenged by the curriculum. While it seemed that she was disconnected from the curriculum, when in fact she had become disconnected from herself.

She completed medical school, but shunned practice and went to graduate school in mathematics, where she both reveled and excelled.

Later, we will talk about the difference between perseverance and stubbornness. Vision helps to discern the difference, but vision only comes with personal reflection. She had always liked science. As a young adult, she had been fully exposed to the concept of the prestige and power of physicians. As a teenager, medicine was a strong attractor, pulling her through her secondary school training.

However, in college, her personality and strengths began to tug away from medicine. In being true to her goal set years before, she did not realize that her talents and academic preferences had shifted. By being loyal to gaining the prize, she had been disloyal to herself. Finally, the pain of the distance between her chosen trajectory and what she craved became too great.

"The heart knows what the heart knows."

The Conductor

Andrey Kolmogorov (1903–1987) was perhaps the greatest probabilist of the twentieth century. In 1925, his first paper on probability appeared as an undergraduate. When he completed his doctorate, he had authored eighteen papers on probability and gone on to write a seminal text in probability theory, as well as treatises in the application of probability theory to the fields of epidemiology, propulsion dynamics, and the movement of the planets.

Yet as a young man, he had no real interest in mathematics. Kolmogorov's mother died in childbirth, his aunt struggling to raise him. His teenage years were roiled by the Russian Revolution. His father, a fighter during the revolution, was killed in 1919; his grandfather (from whom he received his name) was a nobleman who ran an illegal printing press.[*] When he finished secondary school, his record solid, he was positioned to apply to prestigious universities.

Instead, he became a railroad conductor.

For several months, he rode the rails in eastern Russia, helping passengers enter and leave the cars, punching their tickets, attending to their wishes and whims.

After this period, he went on to mathematics at the university, beginning his stupendous and meteoric climb, elevating the field of probability to prominence.

Certainly, no one can know what thoughts filled young Kolmogorov's mind before he attended university. But a conductor's task list was nondemanding. With no radios, there was little communication between him and the engineer. The lazy twelve-hour shifts, with the passengers lulled to sleep by the hypnotic rail cadence, traveling through

---

[*] https://mathshistory.st-andrews.ac.uk/Biographies/Kolmogorov/.

the endless undulating landscape, provided plenty of time for a young man to get lost in the click of the tracks, watching the landscape flow by.

It was time that is good for conducive thought and self-reflecting.

Perhaps young aimless Kolmogorov sifted through the experiences of his life, setting aside what was transient and unhelpful, returning to the others, handling the worthwhile experiences, shaping and reshaping them, and thereby molding his future.

Disconnect Reconnect

The key to the character sabbatical is the sequence "disconnect reconnect."

The disconnection is of you from your work. Overworking drives your work activities and work thoughts deeper into you. You can't seem to let them go. They become part of you and may appear to be inseparable from the rest of you.

This connection is pathologic because you lose sight of where your work ends and you begin.

Reprehensible bosses take advantage of this tight work-self connection. "Well, is everyone else working most of the night tonight?" she may ask. "Are you part of the team or not?" The notion that "the team is everything" quickly devolves into "you are nothing without the team," a pernicious thought poison that you must resolutely resist.

A one-day character sabbatical permits you the opportunity to completely separate from work. You don't have to walk the land in sackcloth and ashes eating locusts and honey, but you should get away from the trappings of your life. To first empty your mind. Then to enter a time of thoughtful, helpful contemplation.

To most of us, the notion of a day sabbatical seems, well, far-fetched. You have a critical work schedule. You have a family. You have friends whom you would like to see and would like to see you. Each of these is also important. Since you face these pressures each week and weekend, the concept of a character sabbatical is pushed farther into the future until it disappears.

> **Character sabbaticals that allow you to extract yourself from your environment to actively consider your station in life are among the simplest and most illuminating times in your life.**

However, ask yourself, if you were suddenly diagnosed with terminal illness, would you not break away from your daily life to regroup? How would you sustain the death of a loved one? By commuting to work? How about the destruction of your marriage? By drinking a beer and watching a Sunday football game with your friends while checking your email as usual? Are these not pivotal episodes in your life when you must disconnect from the daily flotsam?

Don't you deserve time to unplug from "the real world" to reassess, feel pain, develop new thoughts?

In fact, there are pivotal decisions that you are in the process of formulating right now. "Am I going to stay at this institute or not?" That requires important thought, more than just the time that you are stopped at red lights on the way back and forth to work. You may be involved in an issue involving morality at work. Doesn't that require more time and consideration than throwing an occasional thought its way when you are feeding your dog? "Will accepting his marriage proposal lead to my working longer or harder? So should I set aside the notion of tenure and promotion?" These are monumental issues that you are either dealing or will deal with.

Without explicitly generating the mental space to focus on our motivations and values, you can lose sight of what is fundamentally important to you. Or worse, your values and motivations can slowly, imperceptibly alter over time with neither your appreciation nor approval. The midcareer scientist who comes home one night overwhelmed with disgust for her life and accomplishments is a common outcome of this neglect, her values and motivations, under external pressure for years, having shifted.

First and foremost, you require your own attention. Take your sabbatical.

If your significant other is traveling with your children, rather than take the time to research buying a new car or reviewing a recent trove of manuscripts on a particular topic, break away for a character sabbatical for your own good.

Finally, if you cannot take an extended time for such a break, take shorter, regular, more frequent ones. When I would travel for business, I would at my own expense extend the travel by a day. That day, in a strange town, I was alone with time for reflection on the issues I was facing or on what I had just been a part of. To be open to myself. To let me think the thoughts I needed to think but had been too busy to take the time for them.

The "lost time" of all these days was handsomely rewarded as I rediscovered and adjusted my career over the years.

Given the amount of travel I did,* this time of reflection was the best investment in myself that I could have made.

When my travel schedule finally slowed, I would take weekend drives, leaving before dawn and driving for five to six hours.** I was still available for my family by late morning, but would have spent twelve hours in contemplation over two days.

This is the calendar fight for your character-sabbatical time. But once you have won, and the time is on your hands, what do you do with it?

---

* Sometimes thrice-weekly round trips from Houston to Washington, DC.

** I don't know how I would have survived medical school without them. There was too much to absorb. I had to fight for the time to absorb it the right way.

5

Empty Your Mind

War

Make no mistake. You will face internal resistance to this notion of a sabbatical.

Your mind loves a fight, especially an emotional one. It will rail at you with thoughts and ideations to banish this "sabbatical nonsense" and get you back to "the real work," that is, to listen to it. It will argue that it has been your guardian and protector, saving you from crisis after crisis your entire life. Are you going to ignore it now?

Your mind is spoiling for a brawl over this, and therein lies the key.

Don't fight back. In fact, don't argue at all.

An Experiment

If you have never vacated your mind before, then you are in for a treat.

We have all experienced the absence of thought. Suppose you just heard that your significant other has been in a terrible accident. Your son has not arrived at his summer camp yet. You won $300 million in the lottery. You walked in on the intimacies of your significant other with someone else. Your sixteen-year-old is pregnant. The state police show up unexpectedly at work and ask for you.

Shock. Numb. Nothing to think about.*

---

* You might ask yourself why, with your mind bombarding you with thoughts every day, all day, it just shuts down when you might need it the most.

These admittedly extreme examples demonstrate the possibility of emptying your mind.

Our goal is to conduct this vacating voluntarily for our own betterment. So let's do it. Right now.

Look up from this book and ask yourself, "What am I thinking?"

The answer is in itself an illumination. You're thinking lots of things, right? The noise outside. What someone told you at work. The manuscripts that you still need to read. That terrible email from the colleague you don't like. How awful your new computer is. Do I really have a trip tomorrow? I don't know where my car keys are. Will there be a negative editorial about my paper? Why is the room always so dim? Where is my girlfriend? I don't like this apartment.

It can also seem that you can't stop thinking these thoughts. Your mind can be like not just one, but five or ten strong misbehaving dogs, tugging hard at their leashes that you hold, pulling you hard left, then a sharp tug backward, then suddenly all the way forward all at once. They don't stop, and they are always there. It appears to be both random and forceful, and you are attached, like it or not.

Good Servant, Bad Master

Your mind can be a wild, raging thought river, uncontrolled strong currents smashing and colliding, amplifying and intensifying, with you caught up in it all. A maelstrom. Always swirling, always powering forward.

If you thought just a few pages before that a character sabbatical would be unproductive, then ask yourself how "productive" your mind is left to its own. It's like a tornado. Tearing through flotsam thoughts. Making more flotsam. Tearing through that.

This short character sabbatical has already earned its pay if it has opened your eyes to the clutter of your mind. Your mind is like fire—a good servant, but a bad master.

Managing the Dogs

So let's deal with these dogs. Take the first thought off of the stack of thoughts that is plaguing you. Don't pick and choose. Just take one.

Now, inspect it. Make sure you understand what it contains and what your mind wants you to do with it. Then decide, "No, I reject that thought. I won't be dwelling on that thought now. So what else do you have for me?"

> **Your mind is like first a good servant, but a bad master. Start letting your mind serve you.**

That's it.

This is akin to letting one of the dog leashes go. Let the dog run off. Where is it going to go? Somebody else's mind? Ultimately, it will come back. But when it does, you only pick up the leash when you are ready.

Put no energy into the rejection. This is not a fight. Don't turn your thought repudiation into a "battle for 'mind control.'" Our minds love to fight for thoughts, to stoke them up and then ram them home, supercharging our emotions. For example, the agitation that the thought "Aren't Amanda's children prettier than ours?" churns up is already emotion laden. "How did I start thinking about that?" you ask. It saps your strength before you realize that you are immersed in the thought emotion.

If your mind is the raging river, then stand up, climb to the riverbank, sit, and watch the waves hurdle by. Watch their turbulence, being thankful that for at least once, you are not caught up in it.

You have separated yourself from your mind.

You are no more your mind than you are your triceps or your pancreas. Sure, you need them, but there is much more to you than them. Same with your mind. It is a tool that you own and use. It is not you.

Don't Blame Your Mind

Don't get angry with yourself or your mind for "undermining you" or "betraying you" with its self-inflicted thought chaos.

It's not that your mind dislikes you, or wants to psychologically injure you, or hates you. Nothing like that.

It is simply that your mind doesn't know better.

Serving up thoughts is what it does. It's not malicious. It is simply all that your normal mind knows to do. It thinks that pushing thoughts to you is necessary for your survival.

Like a jackhammer that is locked in the On position, making noise and blasting through thought concrete is all that your mind knows to do.

But now, things are different.

You are now the governor of your mind.

So you use your mind like you use an implement. Like a car, a wrench, a trackball, a remote control, or fire, your mind is a tool at your disposal. Many times you use it.

Sometimes you simply put it down.

And to empty your mind, you simply take each and every thought that it offers to you, inspect it, then set it aside, and wait for the next to be set aside as well until no more are left.

Stubborn Thoughts

Some thoughts don't want to be banished. Suppose your first thought is "Is my daughter on drugs?" A powerful, primal thought. It acts as though its undeniable importance overrides any energy you use to push it away. It demands your full attention. It keeps coming back.

However, the principal rule is that it is now you who govern your mind. Not the reverse.

All that you need to think in response, quietly, clearly, is "I am not dwelling on that thought right now. What's the next one?"

Should it come back, and it likely will, don't get angry, don't get frustrated, don't reject yourself. Simply say again, "No, not now." Act like you are talking to a well-meaning but persistent employee. You will get back to him or her in time.

I have had thoughts (imperative thoughts that required careful consideration) come back at me ten times. Fifteen times. Each time, I conduct the emptying process. As long as I am insistent and persistent, they ultimately recede. I know that there will be times when I think about them, but that time is not now.

Congratulations

Congratulations, you are now governing your mind. It may be the first time in your life.

You are not "controlling your mind." That would be a battle that your mind would love to join. You are also not suppressing anything. In fact, you will consider and inspect any and every thought that your mind throws at you. You are not stopping your mind from doing anything. Let the thought stream flow, as it does naturally. You are simply letting it know that for now, you will not be dwelling on thoughts. You are out of the stream. And from this point forward, you will be dwelling on a thought only if you affirmatively choose.

Like any new skill, this requires practice. I try to do this every day. I don't have to be alone. I can be on a plane; I can be sitting in a noisy terminal or a quiet tub. Though banishment with the view to a character sabbatical is a relieving, relaxing skill.

Alternatively, if I have not practiced "mind relief" for a week (for example, being in the midst of an interstate move), it can take me two hours to empty my flotsam detail-filled mind. I much prefer to do it each day.

Doing this each day is a great exercise. When you first wake up and your mind "hits you" with the agenda for the day, you have a wonderful opportunity to fulfill your governor's role, inspecting and rejecting each of them (after all, do you really have to resolve who is picking up your mother-in-law that night now at 3:47 AM?). Each day, you become a stronger governor, and your mind-emptying exercise becomes much easier, even natural.

When you get to the point that the thoughts all but ask permission to be thought when you can inspect them before they capture you and your attention, you have arrived.

You are deciding what you think, not your mind. And in the process, you have given yourself clutter-free space.

What to Do with This New Space

To many people, the vacated mind is the goal. They are now free to be at complete peace with themselves.

Some of your most creative ideas can "come to you" from here. This is the root source of innovation, the "eureka" moment.

So enjoy the thought-free space. You have earned. it.

However, our goal is to embark on a character sabbatical, so how do we use this space to build character?

6

Using Your Sabbatical

Remember when you were emptying your mind of thoughts in the last chapter? Did you find any thoughts concerning courage? Did any of them inform you about your willingness to self-sacrifice? Did you learn about your ethical position as you reviewed and then dismissed the flotsam?

Chances are that the answer is no.

It is a no because these issues of courage, vision, perseverance, self-sacrifice, and moral excellence, the hallmarks of professionalism, are not just passing ideas in the river of the thought chaos—they measure deeper depths. They are the anchors that determine and set your personality and character. You will not find them in the daily detritus of your mind.

A poet might say that these tectonic forces are found in your heart. I would not deny this profundity, but simply say that the way to access them is through a mind that is emptied of thought debris. This empty space that you have generated created a clear tunnel to who you are.

And this undeniable need for you to understand who you are flows both ways. The forces and components of your character need access to you in order to influence your thoughts, your words, and your actions. This access is best *sans* rubble.

So access to your core is not just for you to see who you are. It is also to permit your character and personality—your spirit—to wash over you, ensuring and strengthening their connection to you and what you show the world.

Character and Core Principles

Core values are those of self-value, courage, compassion and empathy, self-sacrifice, morality, work ethic. Your combination of strengths and weaknesses among these core principles comprises your character.

Each of these issues requires focus and attention during the sabbatical. For example, although you believe that you are courageous, you may find that you are troubled because you did not stand up for a colleague who was unfairly criticized by other team members. Was there something special about this circumstance, or were you backsliding away from who you are?

An empty mind permits you to examine yourself in these fundamental areas. You will have the answer before you finish asking the question.

Why you have fallen short or exceeded your self-expectation is worthy fodder for the character sabbatical. How you will conduct yourself differently in a future scenario is informed by a reconnection of you to your core principles. You have the perfect opportunity for this affirmation of reconnection during your character sabbatical.

And there is no better way to start than with checking your own standing.

Self-Valuation

Self-assessments for some of us can be mercilessly destructive. They can tear us down, even bring us to tears.

So no self-assessment should begin without first opening yourself up to a self-valuation, ensuring that your self-respect is intact. Your character sabbatical provides the space for you to effect self-respect healing.

Many young scientists throw their hands up in frustration when they listen to or read about self-respect. "Why should I spend any time ensuring that my self-respect is intact? I'm fine," they assert.

The answer is simple. You need to pay attention and strengthen it because your self-value is continually under assault.

For example, you work in an environment that can be consumed with negativity. Or the job that you have coveted hasn't gone through for you. A grant goes to a rival research team. Your plans for promotion go awry. Manuscripts may never get published. A colleague appears to go out of their way to criticize you.

You are not a robot. You are porous, and these circumstances and criticisms can generate self-destructive emotions in you, for example, that you are not worthy.

They have to be counteracted at once. The simplest and easiest antidote is the preexisting belief that you have value separate and apart from your circumstances. This is a natural product and benefit from character sabbatical and the resulting character growth. This is self-valuation and the beginning of self-respect. You must always check that this is intact because it is always under attack.

Self-Respect

Self-respect is your recognition that you possess an innate high value separate and apart from your weaknesses, performance, or your assessment by others.

Self-respect is not the respect of your accomplishments; it is the respect of the person who produced those accomplishments.

This is a powerful concept. Acknowledging your high value in the face of and despite your productivity and your weaknesses allows you to address your weaknesses from a position of strength that is directly derived from your sense of worth.

A clear demonstration of your self-respect is your willingness to shift your concentration from your daily productivity/task list to your own character and self-development. This "attention to self" is a first affirmative step toward converting these weaknesses into your strengths.

Always be ready to reaffirm your value.

As an example, I boarded a plane from Houston, Texas, to Washington, DC. After stowing my luggage and sitting in an aisle where the other two seats were taken, I pulled out my phone and scanned my email once more.

There it was.

An email concerning my manuscript. There are many manuscripts you contribute to and feel enthusiasm for, but there are few where you are espousing your own belief, your own model, your own mathematics. These are solely yours.

This was one of them.

It had been rejected.

By five different journals.

At this point beyond the midpoint of my career, I was steeled to rejection. The intensity of manuscript competition vacated any notion of early acceptance. One is first rejected before acceptance just as one first fails before they succeed.

But this time, I had let my guard down. I really ached. Not one, but five rejections. All expressing not simple concerns about the manuscript's techniques, but (and worse) utter lack of interest. What was the purpose of all this? At this point in my career, I felt the clock ticking. Hours directed one way cannot be regained. What was I actually doing now that mattered?

And then a thought.

Watching people board the plane engrossed in conversations, struggling with bulky overheads, haggling over seat assignments, I asked myself, "How many of these men and women are stroke-free because of my work?"

Of course, I didn't know, but it was a good question, so I pursued it. How many of them were also heart attack–free? Were heart failure–free? Diabetes mellitus free? I have conducted clinical work in each of these fields, leading to medicines and maneuvers and practices that reduce morbidity and mortality in these fields.

These passengers did not even know me, but I and my colleagues had improved their lives. The force of health care produces a tangible benefit for which we researchers were the anonymous triggers.

That was something to feel good about and a step toward rebuilding self-respect.

Competence and Survival versus Prosperity

You are competent in a challenging technical field. Reflecting on your training and motivation, it is your goal to wield this competence skillfully. To covert education, knowledge, and training into expertise and productivity.

Spiritless competence is eventually ground down by the unremittingly relentlessness of daily problems. Choose to be powered by an attitude, to be propelled by a belief.

Why Are You in Science, Anyway?

It is useful to begin with two acknowledgments. The first is that you're uncommonly intelligent. Nothing about the current frenzy of activities that characterizes your early research career has or will ever change that. Even though you may not always feel smart, and sometimes you may think that you do not act like you are smart, these ephemeral situations are uncommon. One of your enduring characteristics is your keen mind.

Yet despite their cerebral talents, most scientists do not wind up in the upper financial crust of our society, and many of your nonscientific peers, who are not as intelligent as you, will make more money.

Why then would an astute person like you choose a field that won't make you rich?

Lives of Others

Let's begin by recognizing that your unique combination of scientific precision, talent of perception, technical skill, facility with numbers, and sense of charity holds the great potential of profoundly affecting the lives of others.

Unlike an investment banker, you cannot, with a few well-considered keystrokes, earn $475,000 in a single morning. However, your work can beneficially impact a population of thousands, tens of thousands, or millions of people; the pecuniary manager cannot come close to having this impact.

A single well-designed and well-executed research effort can change the paradigm of health care for a critical need.* Consider the work of James Lind [1] who, over six days on a ship in 1747, completed work that saved the lives of thousands of sailors who undoubtedly would have fallen victim to the ravages of scurvy. The link between fine research and beneficial population effects has been the central theme in most research for over two thousand years.

You yourself are involved in the design, execution, and analysis of research efforts whose effects can be seen in people of all economic strata, creeds, and colors. If you're a junior educator, then you'll provide the training that others need to carry out the analysis and interpretation of these research efforts.

Therefore, through an intricate and unique combination of intelligence, determination, altruism, perception, and strength that only you possess, you have started a career that will positively affect the lives of people and families.

Essentially, you have chosen to focus on societal improvement and not financial excess. People don't wait for the arrival of yet one more millionaire. However, they anticipate and will rely on your judgment, compassion, productivity, and vision to help them improve their own lives.

---

* The SAVE, CARE, HOPE and ALLHAT clinical trials are fine modern examples of such clinical studies that have produced wide-ranging beneficial effects on the prevention and treatment of ischemic heart disease.

Inadequacy of the Survival Mentality

Yet however lofty your mission is, your feet are still stuck in the muck of getting through difficult days.

You need some coping skills.

Start with an attitude adjustment. Shun the survival mentality.

Survival is simply the state of being alive. Since as a junior scientist you have too much to do than you possibly can, it is all too easy to set your sights on "just surviving," that is, completing as many of the required tasks that lay before you. All of your skills are devoted to "getting through the day" and completing your task list. Just surviving can seem like quite an accomplishment given the chaotic days that a junior scientist experiences.

That is precisely the problem with the survival mentality.

Surviving implies that all of your talents and energies are expended on simply "getting through." However, if all of your resources are committed to just "making it until the end of the day," then you have no energy for your own development. In addition, "just surviving" implies that by consuming all of your resources, you can provide no help to anybody else.

This is a poor start on the path of service for the good of others.

"Survivalism" is a minimalist goal that you as a scientist must resolutely resist.

Replace it with the notion of prosperity. Prosperity subsumes survival, incorporating not just existence, but also the development of strength in reserve.

Prosperity for you as a scientist means that having worked patiently and diligently to assemble a good career trajectory for the future, you're comfortable with your current location on that trajectory, knowing that your condition will improve as time and your consistent efforts move you forward on that path to your ultimate goal.

Prosperity is being comfortable and at peace with where you are rather than restlessly and ceaselessly agitating for what you don't have. The prosperous person has resources not just for themselves, but also provides time, encouragement, and support for others.

Prosperity is not an automatic attribute of the talented. It must be affirmatively sought.

Replacing the agitating passion of superheated days with the peace of prosperity pays immediate dividends to you.

First, you're free to deviate from your daily task list without guilt because you understand that you'll find the time and resources to handle whatever unexpected issues arise. A prosperous scientist or faculty member can dismiss (or think above) the day's task list since neither triumph nor catastrophe arrives on a preprogrammed schedule.

Expect and anticipate that your day will be both more rewarding and more challenging than what your list of daily scheduled activities suggests. Also, understand that you'll have the strength and insight that you need to meet those challenges, even if you cannot find those characteristics in yourself at the moment. Rest assured that the strengths that you'll require will be within easy reach when you need them.

> **Prosperity means that having assembled a good career trajectory for the future, you're comfortable with your current location on that trajectory, knowing that time and your consistent efforts will propel you forward.**

---

1   Gehan, E.A., Lemak N.A. (1994). *Statistical in Medical Research: Developments in Clinical Trials.* New York. Plenum Publishing Company.

7

Character Growth

My brother used to tell me that one of the advantages of leaving your hometown to go to college or to find a job is that you have the opportunity to reinvent yourself.

No one knows you in the new town. You are free and unencumbered by the past. You can develop new likes. Explore new foods and music. Dress differently. Choose or lose a nickname. You can try new things because the link to most old customs and habits has been severed.

This is actually available to you at any point. Character development is not complicated. Begin your ongoing sabbatical by asking yourself the following two questions:

1. Who am I?
2. Who do I want to be?

Source of Esteem

As a junior investigator, you will have the opportunity to work with astounding researchers with national and international reputations. Although invitations for you to join a research team composed of established investigators can be fulfilling, this soil of appreciation can grow the thorns of discontent.

Specifically, if you don't carefully self-monitor and self-calibrate your response, you can be lured into placing your own sense of self-worth into the hands of your new collaborators. While this can be satisfying when they hold you in high regard, it will almost surely damage you in the long run.

One reason that you're particularly vulnerable to the temptation of turning your sense of self-worth over to others is that at this early stage of your vocation, they appear to hold your career advancement in their hands.

For example, whether you obtain an increase in salary is a decision decided by others. Your suitability for promotion is based on what others think of you. Whether you win an award for achievement is determined by someone else.

Based on these observations, it seems that a useful calibration of how well you're doing is how you're perceived by others. It is this natural metric that junior scientists commonly bring to their first collaborative effort and represents an important issue of character and confidence that must be faced before any real collaborative effort should be initiated.

The problems with turning over self-worth may not be apparent at first when the project is proceeding well and the interactions between you and your colleagues on the research team are smooth.

However, when difficult decisions have to be made, and your point of view diverges from those of your more established team members, you may no longer be held in such high regard. Since you disagree with other team members, they may communicate to you that they don't value your opinions quite so highly anymore. Perhaps word gets back to you that they no longer value your participation. This can create an important personal dilemma for you.

The problem associated with putting your self-worth in the hands of others is that while their approval buoys your sense of self-value, their criticism can capsize it.

This occurs when you make the mistake of translating these disparaging comments into the belief that since they do not value you, you should not value yourself.

The false sense that you should punish yourself saddles you with one of the worst possible and least-deserved castigations—self-rejection.

Self-rejection is a first step to intellectual and emotional self-destruction that you must resolutely resist. While you can afford to lose many technical arguments that deal with the scientific issues at hand, you must not lose the fight for your own self-esteem.

> **While you can afford to lose many technical arguments that deal with the scientific issues at hand, you must not lose the fight for your own self-esteem.**

Self-esteem is the core belief that you have innate value independent of the opinions that others have of you and that you're commendable regardless of what the outside world thinks of you. Sustaining self-esteem is critical to your professional development. As you gain experience in your field, you'll make many administrative and scientific decisions, decisions that will generate criticism and disapproval. This disapproval can profoundly damage you if you let it affect your self-esteem.

Self-esteem is not the blinding belief that you're always right. It is the conviction that you retain your value even when you're wrong.

Thus, solid self-esteem does not inure, but instead opens you to the comments and criticisms of others. Separating and shielding your sense of self-worth and value from these critical comments insulates and protects you. In fact, this separation permits you to listen to, consider, and accept criticism without injuring your core self-value and begin your self-destruction.

Working in a research effort without self-esteem is a painful affair. The unfortunate junior scientist who takes this tack works hard for the approval of others because that approval is linked to her own self-value. However, her colleagues, being human and imperfect, sometimes provide inaccurate and unreliable feedback to her. In the face of this destructive feedback, this junior investigator may compound this error by diminishing what she thinks of herself.

With her self-esteem reduced, the junior scientist becomes dysfunctional as a collaborator. Lacking self-confidence, she is no longer able to trust her intuition and insight, and the progress of her scientific work slowly grinds to a halt. When her research team needs her in a crisis, she is unable to exert her influence productively. The self-visualization that she is valueless and can provide nothing of consequence sinks her performance much like a heavy leg weight sinks a swimmer.

Self-worth is a natural need. Therefore, the scientist who links their own self-worth to the opinion of their superiors will do almost anything to gain their approval. Sometimes this weakness is deliberately exploited. A manipulative senior scientist, recognizing how susceptible you are in this sensitive area, can attempt to control you. Essentially, they can dangle acceptance before you, enticing you to carry out work or analyses that you know are wrong. Nevertheless, you may choose to carry out this work because its execution will bring approval and, with it, a false sense of self-worth.

A statistician who worked for a private corporation fell into this ruinous problem. He was an accomplished scientist who functioned effectively as a team statistician for many years. Always anxious to please his bosses and superiors, he was intimately involved in the design and analyses of many productive research projects.

However, his ability to contribute was inexorably overtaken by the undesirable driving urge to keep his bosses happy. Rather than work to persuade his superiors that his innovative approaches to important technical problems were sound and worthy of consideration, he instead prematurely ceased defending his own professional point of view in the face of his bosses' displeasure.

The ultimate result in this case was tragic. Being unable to sustain his self-value in the face of criticism from his superiors, this promising researcher lost all confidence in himself as a competent scientist. With no self-assurance, he no longer trusted the results of his own calculations. Unable to find his own talents within himself, he was reduced to sending every computation that he was asked to execute out to other statisticians, asking that they confirm his arithmetic. After working for many months in this wretched state, he left the industry altogether.

It's especially noteworthy that low self-esteem was held by a senior investigator, begging the question, "If a senior scientist is warped by this environment, what hope does a junior colleague have?"

The answer is the junior scientist must begin at once to (1) recognize that the healthy development of her sense of self-worth, separate and apart from the criticisms of others, is essential to withstand the pressure that inevitably will be brought to bear against them at some point in their career, and 2) there will be circumstances where she will have to choose between allegiance to their own principles with attendant serious consequences or the abandonment of those principles to the demands of their superiors.

This is the beginning of courage. One of the important advantages of early investment in character sabbatical and healthy self-worth development is the development of this strength of heart.

Before you proceed with your collaborations, first ensure that you can readily identify your source of self-worth. If that source is derived from a well-anchored sense of your value and significance that is independent of external events, then proceed with confidence into your new environment.

However, as for most of us, if you're uncertain of your source of self-worth, stop first and examine it, with the view of testing and repairing it. Specifically, sever the link between your self-worth and your acceptance by others. Use your character sabbatical to develop a solid, internal, and unwavering source for your sense of value. This source should sustain you regardless of the circumstances of your research.

> **There will be circumstances where you must choose between allegiance to their own principles with attendant serious consequences or the abandonment of those dear principles to the demands of your superiors.**

Character growth develops for you a solid sense of self-worth that in turn provides the stability of your bearing in a tempestuous intellectual environment. The ballast of an independent sense of self-worth helps you to maintain your balance in the face of these unpredictable interactions.

Recognize that as you progress and your character developments, your sense of self-worth will come under assault. Anticipate and expect these attacks, turning them to your advantage. View your most challenging times as the days when your sense of self-value is tested. Use these times to observe how well your source of self-worth sustains you.

Examine and either repudiate or repair your source of self-value if it fails you, thereby shaping your character growth and development. Your short-term goal is to have your sense of worth independent of external criticism. Ultimately, you want it to be separate and apart from your career trajectory.

Working toward this goal should not encourage inconsiderate actions on your part. While you should always consider what others think, do not let what they express influence your deep-seated sense of how you value yourself. Making a mistake should not change your self-esteem, and the need for an apology should not diminish it. In fact, a secure sense of self-worth will make it easier to hear clearly, consider carefully, and apologize freely and openly; while you may be under attack, that attack cannot damage your sense of purpose and self-esteem.

Your appreciation of your own value, coupled with your complete and unconditional acceptance of yourself, is the basis of your ability to successfully and wholeheartedly participate in a collaborative effort.

Moral Excellence

Moral excellence in science is not a passive, inactive, or static state. It is not a commodity that is on the upper shelf in the kitchen, rarely used, like a precious spice.

Challenges such as private sector interaction, incorporation, academic promotion, competition for grants, controversial research lines, private consultation activity, and expert testimony will find you as you advance in your field.

For each of these, you need to choose whether you will take part, and if so, the tenor of your participation. It is all too easy to make innocent mistakes with tragic consequences. A finely attuned moral compass to which you are connected will help you determine if taking part in these activities is right for you.

However, essential instruments require frequent inspection and recalibration. Ethics requires attention and consideration. You may be confronted with a circumstance that you haven't considered before. Test your moral compass by affirmatively using it. Be alert for ethical opportunities while remaining vigilant for the development of any seeds of unethical conduct developing in your own actions. This requires both time and mental space. A perfect topic for your character sabbatical.

Ethical conduct is not the passive passenger sleeping comfortably in the backseat while you drive to your career destination—it is the driver. Make sure that you know where it's taking you.*

Self-Sacrifice 1: General Comments

Productivity is not the leading force in your career, but only one of several important cores of professionalism. Accepting this principle leads to the inevitable, illuminating, and liberating corollary that there are times when productivity should be deemphasized.

The recognition of productivity's importance does not make you its slave. Many junior scientists, eager to start off on the "right foot," create an environment that maximizes productivity, permitting them to mass-produce experimental reports, manuscripts, and presentations.

---

* This is discussed in detail in chapter 16.

The difficulty with this unbalanced approach is its unsustainability. As long workdays stretch into the evening, family life suffers. As five-day workweeks stretch to six- and seven-day work marathons, the natural leisure of a restful weekend is squeezed out. Holiday periods are compressed, and the restoration of vacation is squandered by constantly checking email and calling your colleagues for updates. Thus, productivity is maximized at the expense of personal development.*

The "productivity at all costs" environment is degrading as your personal damage grows and the self-sacrifice becomes the driving force of self-destruction. Ironically, the environment created to maximize productivity ultimately poisons the plant it was designed to support.

Alternatively, build an environment and a work style that focuses on creating and energizing your attitude and sharpening your talent. Specifically, this means first creating the internal attitude then the external environment that allows you to bring the best of your talents to bear on your research issues.

Take the time to develop your best judgment. Take the time to create good standards. Take the time to generate a finely tuned ethic. Each of these comes from careful thought, reading, and consideration during your character sabbatical.

Take the time to develop your best judgment. Take the time to create good standards. Take the time to generate a finely tuned ethic. Each of these come from careful thought, reading, and consideration.

A useful metric is asking yourself when working: "Am I giving the best effort, attention, insight, and intellectual force to this?" If you can't do that because you too tired, or too hungry, or too angry, or emotionally spent, then you have no business trying to write a grant, finish a presentation, or create a manuscript. Fix the attitude problem first, then return to work.

---

* Unfortunately, there are still some bosses and division chiefs who are perfectly happy with your uncontrolled productivity. Indulging themselves in the self-serving notion that "if you're willing to kill yourself, why should they stop you" is the snide justification that permits them to profit from your self-destructive work effort.

Building this attitude and environment takes consistent effort, requiring you to take a productivity pause while you instead focus on your actions, motivation, and values. Consistent effort here sharpens your skills, permits the best application of those skills, and thereby generates the best work from you.

In addition, this atmosphere, unlike the previous environment, is actually good for you. The fact that you have chosen not to sacrifice your best nature to productivity, but instead are willing to sacrifice productivity so that you can use the best of your talents to be productive reaffirms the natural authority that you have over your life. You aren't being productive to gain value. You are productive because you have value.

> **You aren't being productive to gain value. You are productive because you have value.**

Self-Sacrifice 2: Sacrificing Time for Another

Scientific progress is collaborative. This does not mean that everyone works separate and apart from each other. Collaborative is interactive. At its core, collaboration means that you sacrifice some of your productivity time for another.

Commonly, this level of sacrifice is small. You may be working against an imminent grant deadline when you are interrupted by a colleague who needs your help. Perhaps it is a technical matter residing in your wheelhouse of expertise. It might be an ethical issue that they observed about their boss. Perhaps your colleagues need to talk through an insecurity about a difficult presentation with you, requiring you to be a sounding board for a few minutes.

There is no question but that you "do not have the time for this." If you take time from your productivity to help your colleague, you miss your own productivity goal.

I have felt this way often, but have rejected the concept that the time I give to them is time that I lose. This is because I have never experienced, despite fears to the contrary, my self-sacrificing effort for another to come back against me. The entire concept that time is a zero-sum game when it comes to self-sacrifice, I have found, is a theoretical canard, a self-protective idea from my mind.

One of the assumptions in the "I have no time for you" philosophy is that it is more important that you meet your goal than she meets hers.

That may be wholly wrong. It is quite possible that her work stands to make a greater contribution than yours.

This is a turn of the question that we commonly do not consider—that in the end, our colleagues stand to make a greater contribution to science than we do. It forces us to ask why we are doing what we do. For science? Or for ourselves? Sometimes we cannot even know that until the end.

Gravity

Consider one of the important scientific questions of the seventeenth- and eighteenth-century England. What is the nature of gravity, and what laws govern it?

Many scientists and philosophers at the time claimed to have the answer, but upon closer inspection, none seemed to have the proof demonstrating that answer's veracity.

Robert Hooke was one of these. He was a scientist of great reputation, having made stunning advances in the early microscope and, using this new microscope, discovered the cell. Hooke asserted that he knew the law that governs the attractions of bodies for each other and in fact demanded he be given credit for the solution to the problem. However, he never provided the answer or the proof of his answer, demanding that his reputation required that his word be accepted.

One of his inquisitors, Edmund Halley, in frustration went to Cambridge to visit Isaac Newton to get his perspective. Isaac Newton during this period of life was a morose and brooding recluse.

When Halley asked him about gravitational law, Newton glibly recited the law of gravity. But when Halley challenged him for a proof, Newton was unable to find it. Halley was crestfallen, but Newton, unshaken, said that he would simply rederive it and asked Halley to return.

At the appointed date and time, Newton provided the detailed mathematical proof.

Halley, stunned, asked and cajoled Newton to publish this at once. But Newton was adamant. He had had a run-in with Hooke years ago when Hooke falsely accused Newton of plagiarism of his work on the light spectrum. Newton, young, alone, isolated, and humiliated, retreated to his home, vowing never to publish again. He spent his time alone working in alchemy and brooding, applying mathematics to the New Testament to predict the Second Coming of Jesus.

Now Halley left his own work behind, devoting himself to Newton. He spent days with the recluse, finally convincing him to put his writings in one body of work. Halley edited Newton's work, debated with him, and ultimately befriended him. He drew no salary for the effort and let his own work in mapping the stars and a lucrative sea salvage operation lapse.

Finally, Newton was done. The final work was published under the title of *Principia*.

Principia is considered one of the greatest achievements in the history of science. Not only did it explain the movement of the planets, but it also removed the notion that astronomical forces were solely in the purview of God and beyond man's understanding. Thus, the entire field of astronomy was now laid bare, available not just for awe and inspiration but also for study and deduction.

And it would not have occurred had Edmond Halley not sacrificed his time, effort, and what little money he had to be the friend, counselor, benefactor, and psychologist of Isaac Newton.*

The power of self-sacrifice can be explosive.

Your Strengths and Weaknesses

It's difficult to pick up the right new coping skills without knowing your true strengths and weaknesses. You might begin by using your character sabbatical in an evaluation of your strengths and weaknesses.

Interviewing yourself. What do you enjoy about your work? Why? Which tasks do you recognize as necessary but nevertheless try to shun? Why? Ask probing questions, increasing the drill depth until you get answers.

There are many natural athletes with innate talents whose sports careers collapse. The skilled, mature, and successful athletes are those who have learned that complete reliance on those strengths is the path to failure.

They have instead adopted the philosophy that success requires not just their talents but also their willingness to convert their weaknesses into strengths.

---

* Ironically, what is known as Halley's comet was not discovered by Halley. His contribution was his assertion that comets followed elliptical orbits and that they returned at predictable times.

Thus, the basketball player who is a superb right-handed dribbler learns to become even more adept at handling the ball with his left hand. The baseball star who is a skilled hitter also develops superior baserunning skills.

Developing these new skills is difficult and humbling. Working on a weakness requires the skilled athlete to work as hard as, and to appear as clumsy as, the unskilled player. This can be an ego-bruising experience. However, converting weaknesses to strengths expands the dimension of the athlete's performance. An overwhelming fastball can be a formidable pitch. However, if this is the only pitch in the pitcher's repertoire, then the opposition adapts to it, eventually defeating him. By mastering the curveball, the pitcher converts himself from a good pitcher to a masterful one.

In order to succeed, you have to confront the weaknesses that you fear and convert them to strengths.

Get used to doing what you don't feel like doing in order to be the professional scientist that you want to become.

Get used to doing what you don't feel like doing in order to be the professional scientist that you want to become.

As you take stock of yourself and review your development, you'll inevitably turn to your most recent experience. That may be at a previous job or graduate school life. For you, as for everyone, a candid self-appraisal will reveal both triumphs and failures. Examine both, but avoid the easy path of assuming that past victories alone determine your career future. Past failure can also be a wonderful trainer although it certainly doesn't feel wonderful at the time. Defeats commonly lay the foundation for future victories. Alternatively, previous victories can set the stage for future failure.

The seeds of recent victory produce failure for the careless while the thorns of defeat bloom into success for the perceptive.

Long-Term Vision

What will you be doing twenty years from now?

This question appears absurd in an environment where the vicissitudes of life appear to regularly overturn well-intentioned short-term plans. However, it is precisely in this type of chaotic environment that a long-term goal is most useful.

As a scientist, you understand the importance of having a long-term research goal. Before you carry out an experiment or energetically move forward on a project, you must know how this activity fits within your field. How can its results be integrated into your area's fund of knowledge? Without a good answer to these questions, the research endeavor runs the risk of being purposeless.

Just as it is important to have a larger perspective on your scientific work, it is critical that you develop a long-term vision for yourself.

Given that your career will be characterized not by a dearth of choice but an abundance of opportunity, you need a yardstick for differentiating the opportunities you'll accept from the ones to be set aside. Your long-term goal serves admirably here, allowing you to appraise whether the choice facing you will move you closer to your goal.

So what will you be doing in twenty years? Will you be developing new patentable technology? Will you own a company? Will you ultimately be a dean or a public scientist-politician? Will you be a speaker? A regulator? Will you write?

What would you like to do, and what would be good for you to do?*

The difficulty with long-term career goals is not their demand for your loyalty. Quite the contrary, once you choose a well-considered goal that nicely aligns with your natural talents and abilities, it is easy to stay focused on it. The difficulty lies in actually choosing it.

Selecting a long-term goal takes time and effort and is one more perfect thought complex for you to explore during your character sabbatical. It requires the time for you to gauge your talents and weaknesses, separate hope from reality, weigh the direction of the field, and balance what you would like to do with your financial needs.

Also, if you have a family, then what do your loved ones think and expect of you? Decisions about family will often influence, and are influenced by, the long-term goals of you and your partner. This is good work, but it takes effort and time.

---

* If you cannot decide where you want to be and what you want to be doing, you might reverse the question, asking what you specifically want to avoid.

If after much thoughtful consideration, including conversations with your family, your mentors, and your close friends, you develop a vision, then let it capture your focus. A long-term vision illuminates the path that you should walk. Keeping your long-term goal close at hand gives you a new good metric to measure the role that your current activities play in meeting that goal. Furthermore, the development of a distant goal worthy of your pursuit can simplify many short-term decisions that may be confronting you. Decisions about job opportunities, project options, and research group participation can all be simplified in the presence of a long-term strategy.

Without a long-term goal, you're at risk of having your career caught up in the random eddies of opportunity, distraction, and the vicissitudes of life. Together, these forces can sweep you up and deposit you on a shore that you may not like, but from which there is no return.

If you have such a goal, then during your sabbatical, critically reexamine it.* Are you still firmly committed to it? Have you learned anything this past year that requires you to modify your plans or your long-term schedule for executing those plans? How has the mixture of successes and failures that punctuated this last year affected your strategy? Must your goal be fundamentally altered in the face of new circumstances, or has its arrival been hastened? What new activities have occurred in your career that have become distractions that you must shed?

Develop and then regularly challenge your vision with an open heart. Treat the selection and modification of the goal as a serious life enterprise itself, requiring the best of your talents and contemplative abilities. This investment will pay handsome dividends for you.

Develop Perseverance

The very fact that you're a scientist attests to your doggedness. Your ability to hammer against obstacles until they give way has helped you to overcome the challenges of classwork, exams, reports, theses, and dissertations.

However, as useful as this skill has been for you, it may need your attention. While the ability to work hard against an obstacle is laudable, what do you do when the impediment doesn't give way?

---

* I have a colleague who annually takes three days at his birthday to both get restored and to critically review his long-term career and financial plans for the future. It is sometimes inconvenient for him to do it, but he has never regretted the decision to stop his work and consider these other matters on a regular basis.

The heart of the answer to this question lies in the distinction between stubbornness and perseverance. The stubborn individual and the perseverant worker are each focused on a goal. They both work hard, and neither is easily diverted by small distractions. Yet of the two qualities, perseverance is preferred.

As a child on a spring day, I commonly had to wait in the parked family car while my parents completed an errand. Sometimes, a house fly entered the car through an open window and, not recognizing its surroundings, would try to exit. However, it frequently would spend all of its time trying to fly through the same closed window. It knew its attempts to escape were failing, but time after time, it knocked itself against that closed window until it exhausted itself and died. This is quintessential stubbornness—visionless effort.

The obstreperous worker fights to break through an obstacle that for reasons that are unclear to him continues to block his path. Like the noncomprehending fly, his inability to understand why his stubborn efforts fail produces not illumination, but slavish repetition.

Perseverance, however, combines persistence and vision. The perseverant worker takes the time to step back and thoughtfully consider the problem. Sometimes repeated exertion in the same direction is required for a solution while at other times, a completely new effort in another dimension is needed.

For the stubborn, overcoming obstacles is the key to success. For the persevering, the key is continued progress toward a long-term goal that has been carefully considered and deemed worthy of consistent, diligent, but not obsessive effort.

Avoid wasting your time on fruitless battles. Don't splinter your effort into activities for multiple goals, none of which places you where thoughtful consideration suggests that you should be. Recognize that there is a time for activities to be prosecuted in earnest and a time for that energy to be redirected.

For example, in 2002, I agreed to lead a team in the development of a database that would serve as the repository of the information from a collection of research activities. These activities were designed to assess the effect of new treatments to reduce the damaging effects of strokes.

Our group traditionally used paper forms to collect data. However, in this circumstance, we decided to develop an online internet application that allowed remote data entry with simultaneous "real-time" quality control in a heightened security environment.

This was a fine objective, but the technology for these systems then was quite new, and neither I nor the programmer knew much about them. We therefore decided to work on this project together. We would have to start from scratch to learn the technology and then, building ourselves up, would assemble the application that we could hopefully deploy.

This was *terra incognita*. Like many new and complicated projects, our first efforts were halting and full of aching frustration. We expended a small fortune of our own money on books, spending many hours at work and at our respective homes on developing the programming skills that we needed.

It took approximately one month of painful work to produce the first elementary web page. Slowly, ideas and concepts became clearer to us, and several months later, we had a tight, functional application that the investigators could safely and securely use remotely.

In the meantime, the two of us had moved from novices to programmers who were confident in showcasing our work; we were now fully capable of utilizing this technology for new research efforts in different fields. For me personally, this had been a particularly exhilarating experience. I had demonstrated to myself that twenty years spent in research, teaching, and statistical writing had dimmed neither my enthusiasm for, nor my aptitude in, computer programming.

However, during a brief break between projects, I found myself plagued by some nagging concerns. I had spent eight months and approximately one thousand hours working on this project. While this work had not been carried out to the exclusion of my other responsibilities, the press of other professional obligations, including grant involvement and administrative activities, was undeniably real.

In addition, my coprogrammer, who, like me, started out as a neophyte in this new field, was now able to progress independently. While at first we relied on each other, she was now able to proceed without my help. In fact, she had now surpassed my skills and was preparing to share her expertise with other programmers who needed to learn this new technology.

This combination of realizations led me to the conclusion that it was time for a change in my path. Since I could justify my continued deep investment in programming with the new projects coming up, it would have been easy for me to stubbornly argue that I should continue to play a central role in the programming for these activities. However, it was time for me to move on.

My season of programming was fun, but was now over. I would not have recognized this if I hadn't chosen to stop for a time of self-evaluation.

8

Courage

Courage, in the end, is the ability and willingness to suffer the pain of your own destruction for what you believe.

Vision, sacrifice. Pain and loss. Self-destruction. These are frightening concepts for adults who have never considered them before. Frankly, if a character sabbatical doesn't help in developing courage, then really, why bother?

If you believe that one person of courage cannot make a difference, that you cannot and will not make a difference and therefore don't need courage, then consider the following story of the man who saved all of our lives.

9

When The Burden Lands

Individuals in Crisis

In 1962, the Cold War between the United States of America and the Union of Soviet Socialist Republics (USSR) caught fire. President Kennedy was elected in 1960, took office in 1961, and promptly fell into a debacle with an ill-advised invasion of Cuba.[*]

This invasion, known as the Bay of Pigs, was a national and international embarrassment for the young president. The invasion's failure, with resultant televised imprisonment of US soldiers and Cuban insurgents, conveyed to the world that the United States was unable to project its power as forceful under this inexperienced president.

First Secretary Khrushchev of the USSR took advantage of this new weakness at a scheduled summit meeting in Vienna, Austria. He denied Kennedy, desperate for an overarching agreement that would allow a peaceful coexistence between the USSR and the United States, such a joint statement. Instead, the first secretary accelerated the buildup of Soviet forces in Eastern Europe. His construction of the Berlin Wall isolated West Berlin from West Germany, pressured NATO, and threatened this alliances' survival.

It was at this time that Khrushchev decided to place nuclear offensive missiles in Cuba.

---

[*] There are many good books, documents, and reportage describing the crises from 1961–62. One is by Michael Dobbs entitled *One Minute to Midnight*.

No one is really clear as to Khrushchev's motives. Some suggest that Khrushchev, like his Presidium now aging, longed for the fervency of the early communist movement whose spirit was now embodied by Castro.

In addition, Khrushchev loved to brag about his nuclear arsenal, arguing that it could destroy Western Europe. When he was told bluntly that the US response would utterly destroy the Soviet Union in response, he began to consider how to even the odds between the US and the USSR by placing nuclear weapons in Cuba, only ninety miles from the United States. This was despite his assurances that he would never place offensive weapons in Cuba. It was also in contravention to USSR policy that such weapons would never be located outside of USSR's territory.

Nevertheless, Soviet missiles were detected by the United States during surveillance satellite flights over Cuba.

The US military was livid. First, it was still humiliated by the Bay of Pigs failure. Second, the new placement of these weapons was a threat to the United States' survival. By this time, international ballistic missiles were a fact of life. However, while no one really knew if a Soviet launch of a nuclear weapon east of the Urals could hit Atlanta, Georgia, nobody denied that the Soviets couldn't miss if they launched from Cuba.

After several days of heated discussion, the United States announced to the world that it would set up a blockade around Cuba, extending eight hundred miles into the Caribbean Sea. The purpose of this blockade was to interdict the twenty-five merchant ships headed from the Soviet Union to Cuba carrying commercial, mercantile, and military products. The goal of the United States was twofold, first to identify and turn back ships containing weapons that were destined to Cuba; the second goal was to convey to Khrushchev that the United States was taking prudent steps to avoid war.

The first secretary's fulminant response called the blockade banditry. He admitted that mercantile Soviet ships could be stopped by the American blockade. However, he would now support his vessels with submarine warships that began to pour into the Caribbean. In addition, Soviet crew staffs in Cuba were ordered that the nuclear weapons already in Cuba be operationalized so they could go to war at a moment's notice.

The stakes were set, but the strategic confrontation crumbled. Unlike today when worldwide communication is instantaneous, chaotic electronic discussion ruled the day in 1962. Portent communiqués required laborious effort. The communication had first to be translated and translated exactly. Then the communication was sent through undersea cables. It then had to be retranslated, confirmed, and communicated up through the chain of bureaucratic leadership. Communication would take not minutes or hours, but days.

In addition, tactical problems were arising at sea as each country struggled with how exactly to control events between the two navies such that nuclear war would not break out by accident. How, for example, could the United States understand the cargo of a Russian ship if the US ship had no Russian-speaking sailor? What if the Russian captain refused to answer the American hale? Would the US then shoot the rudder and propeller off the Russian ship and tow it to Jacksonville? What if once there, the ship's cargo held nothing but baby food?

The Soviet side had its own difficulties. Would the decision to stop a Russian ship by shooting out its propeller and rudder be an act of war? International maritime law said yes. Would Russia then attack with a nuclear barrage against the United States from Cuba and Russia that would kill over 150 million Americans, inviting a return assault that would kill 200 million Russians?

Both sides struggled to begin to answer the myriad of problems that would happen while the fleet commanders struggled to manage the events. It's one thing to engage in war games where these concepts were casually and eruditely discussed. It is quite another matter to consider the same options when thousands of tons of enemy naval hardware are bearing down on you.

And yet the ships converged.

Kennedy, in an attempt to slow the accelerating process to war, contracted the arc of blockade around Cuba from eight hundred miles to five hundred miles. However, this put the blockading force within easy reach of fighter jets in Cuba. Two Russian ships avoided the blockade altogether, turning back off the line toward the USSR. This provided the world some hours of relief until the US Naval Intelligence announced its belief that these two ships were merely moving to group with a new unit of Soviet submarines that would now move toward Cuba to defend them. Steps to avoid war turned into steps back into one.

Both the United States and the Soviet Union had set the stage for the strategic war to be waged. However, neither was prepared for the tactical consequences.

This was the major point of the Cuban missile crisis. Leaders of major countries can control the strategic direction of their forces, but they cannot control the actual tactical operation, which depends on single military commanders and units in the field. It depends on men and women.

Then disaster struck. A single Soviet surface-to-air missile was launched, destroying a US surveillance flight over Cuba.

The US Air Force leadership, furious at this attack, all but assaulted President Kennedy for not moving to attack Cuba and destroy all the Soviet Air defenses. But just as the air force argued that they were sending men up and therefore had an obligation to defend them, Kennedy argued, didn't the Cubans have a right to defend their airspace from a marauding US jet fighter? The vituperative arguments were never ending. Should one junior officer who pushed the button, he shouldn't have, and killed one pilot be permitted to start World War III?

Khrushchev then learned that the United States had launched a U-2 surveillance flight over not Cuba, but the Soviet Union itself.

When the First Secretary heard this, he was furious. However, like Kennedy, he was beginning to think about the consequences of his actions. The surveillance flights were frustrating, but routine to the first secretary. Yet now the Soviet Air Force launched and planned to shoot the US surveillance flight down with nuclear missiles. Khrushchev realized that a successful attack would lead to essentially a nuclear detonation over the Soviet Union involving a US aircraft, leading to World War III. Khrushchev ordered that the surveillance plane be ordered back to the United States with a Soviet escort and not be attacked.

However, while both Khrushchev and Kennedy, recognizing what little control they actually exerted over the totality of their own forces, struggled to gather control of their armed forces in their hands, neither considered that they could not control forces with which they could not communicate.

The sovereign Soviet submarine B-59 had been deep underwater near Cuba for two weeks now. Their last orders: fire against the US blockading forces should the submarine be attacked.

However, those orders now were one week old. With the air in the submarine becoming stale, food becoming rancid, arguments began to break out on the boat as to what they should do. The submarine was essentially incommunicado. The only noise they heard was the steaming of the US destroyers overhead. They believed war had already broken out and that they must launch.

However, Vasili Arkhipov, the brigade chief of staff on the B-59, disagreed. Arguing that they had no current operational orders and that a mistake on their part could make the situation far worse, he was able to hold the other two officers who were in charge of the weapons at bay. Emotions were charged.

Then the US attack on the Soviet submarine began.

Depth charges began to fall, shattering the remnants of the emotional stability of the crew, raising the arguments to screaming matches as to what they had to do. The submarine captain's orders were clear. If you are attacked, you fire back. Yet the weapons he had were nuclear weapons that would certainly demolish the American blockade and escalate the crisis.

They all also knew that Russian officers on submarines do not disobey orders. If those orders are not carried out, they return home where they and their families are sent in disgrace to Russian gulags.

The entire crew and two of the three submarine leaders argued that the orders be followed out and they attack. However, Chief of Staff Arkhipov, the third officer, recognizing now that his career was over, continued to argue against the counterattack for hours.

Suddenly, the depth charging ceased. After several minutes, the captain agreed to surface the submarine. Reaching communication depth, they learned that in fact, the crisis had been resolved that very day.

This courageous officer, Arkhipov, saved the world. If he had not stood his ground, you and I would not be having this conversation.

And what of this officer?

Stories are conflicting. One story says that he and his family were disgraced, forced to live in poverty in Moscow until he died of radiation poisoning. Another story says that his entire family was sent to live out their existence in an eastern Russian gulag where they died.

For those who argue that one person cannot make a difference, this one person saved our lives, yet most of us have never heard of him.

Why Does This Matter to You?

You can make a difference and save the world one day, any day. Connect to your character so you can know how you're going to respond and the degree to which you are willing to sacrifice.

We will say much more about courage. But it begins by testing, focusing, challenging, and questioning your character. What will you stand for, and what will you not?

Answers come from a deep investment in your character sabbatical. Prepare for the day when the burden lands on you.

Will you be ready?

10

Deploying Courage

I want to be up front with you.

This will be a very challenging chapter. If you devote exclusive time to character and the development of courage, you will prosper although your career trajectory may be altered if not shattered.

If you do not, then you will lose yourself and what you stand for.

This is a terrible choice. But, in our demented slaughterhouse of a world, science does not exists in a vacuum. Here, terrible sacrifices from the good among us are required. These are not sacrifices of principle. Quite the contrary. Principles stand; but loss of position, prestige, bonuses, power are sustained.

Courage requires the strength to see your career trajectory changed, even destroyed because you stand for your principles.

There may be some of us (God bless them) born with courage. However, most of us have to grow it on our own. It is in the space of the character sabbatical that you achieve this. Here, bring in a tough ethical situation. Maybe it is local, maybe it is national. Ask yourself what would you do in their position, and most importantly, regardless of the answer, why? Be absolutely honest even though you may not like the answer you get. What answer do you want? For what do you stand? One that requires money, or one that takes something from you? Is the sacrifice of real worth?

Can anybody take your self-value? If not, is anything else taken from you really worth having?

It is essential that you ask these questions and do this development before you are in the midst of a crisis. Now is the time for you to leisurely explore in a stress-free atmosphere your answer, your redline. Embrace a position; stand for an action that you will be able to call your own and hold fast to. In this fashion, when the crisis comes, you have already done the hard work. You know what you're going to do. The decision is easy. And the sacrifice is worth it.

That is courage.

Essentially, there are times in your career when you'll have to choose between two egregious alternatives. Each requires you to pay a different but unbearable price. Regardless of your decision, you lose. Prepare yourself to do what is right, with the conviction that selecting the right alternative is its own reward.

For example, choosing to leave a vituperative and truculent principal investigator requires you to pay a huge and painful price. You may lose any or all of the productivity dividend that would have come from the scientific work that you invested in her group. You are hardworking and, until this episode, were well on the way to a deserved reputation as a reliable team member. You do not want the mark of troublemaker.

However, the wrong choice requires you to pay a price from which you'll not recover. A punishing pattern of abuse permanently damages your ability to function as a healthy collaborative scientist in the long-term.

While you can and should give the best of your efforts to your chief, you must sacrifice your spirit of self-value and good ego strength to no one. Preservation and growth of these irreplaceable character features is what you gain by leaving.

Take Their Place

It is important to avoid second-guessing the performance of others. Spend some time reviewing the following:

Dr. Deborah Birx https://www.youtube.com/watch?v=lFKQGGf1iiI
Elizabeth Cheney https://www.youtube.com/watch?v=4R1Nu23nx2o
Dr. Tony Fauci https://www.youtube.com/watch?v=pFoaBV_cTek

Each of these individuals chose to take a position. Were they courageous in your mind? More importantly, what would you do in their circumstance as they did? Something else? Why? With the exception of

Rep Cheney, neither anticipated they would be in that circumstance that day. Were they prepared or unprepared?

These circumstances will likely not happen to you. But other ones will. Perhaps worse. How do you plan for these? And how do you decide what to do?

Why Do Principles Matter?

This is an elementary question, but let's explore it. After all, if courage requires a sacrifice of great cost, what then balances the cost? Why are principles worth so much?

Principles are the glue that holds society together. Without them, culture falls apart. There is no longer a common bond between individuals. With no societal compact, we lose the common understanding that binds us together or that governs our interactions. We no longer work together. Families themselves become dysfunctional. There is no common connection between individuals that sustains, but only ephemeral common interests. Society with no bond falls apart.

If you want to see a community without principles, go to a dog pound.

Principles have developed slowly, lagging behind the development of technological advances in civilization. Different religions attempted to incorporate them. The Ten Commandments, the Koran, the Magna Carta, the New Testament, the *Rights of Man*, The Declaration of Independence, the US Constitution with its amendments, and the Emancipation Proclamation are epoch moments because they asserted what early civilization was aching toward.

That men had value, innate to themselves.

In their own way, they asserted that bonds between people are based not on mercurial connections but the sustaining bond of value and respect. This concept slowly expanded to finally include people of color, women, and the LGBTQ community. These are still under assault, but human dignity—the innate value of individuals, not because of who they are, but simply *that* they are—is a bedrock of our civilization, nation, state, city, community, and neighborhood.

Without principles, we have nothing together, so we fight to sustain them. And we pay a price for principles because without them, nothing of value is left.

Costs and Their Anticipation

We have talked about costs. Loss of prestige. Reduction in salary. Relegation to a smaller office on a less desirable floor. No access to the top postgraduates. Lucrative consultancies lost. Your career is not just altered, it's shot down. Thus, a smaller house for you and your family, and state schools rather than private schools for your children.

These are the costs of your principles.

If you are not prepared for these staggering losses, you may not be able to sustain them. In fact, you may give up on your principles to retain them.

The key to accepting and then adjusting to these losses is to anticipate them. To expect them. To know that you will have to deal with these important setbacks in the future.

In your character sabbatical, examine your reactions to these potential losses. Examining these issues early permits you the leisure to turn them over and over again in your sabbatical without dealing with the emotion that attends actually sustaining the loss. What does it mean to you to have to downsize a house? What real loss will be sustained? Will you be reduced in some important way? Can you recover? Is it better to have a smaller home in the first place?

Use your sabbatical to grade how important these potential costs are to you. By exploring these issues early in your career, you have the opportunity to arrest their growth in your eyes, to perhaps move from "I need them" to "It would be nice to have them, but they are not all important. I could live without them, as I have so far."

Bring your significant other into these conversations. You are beginning to make a determination about your lifestyle based on your principles. He or she has the right to know.

You have now calibrated and corralled the loss. It is acceptable to you.

And you have addressed the most important question. By devaluing theses potential costs, you are close to the answer to the critical question, "By holding fast to my self-worth and self-value, what can be taken from me that really matters?"

11

General Tools and Skills

After you have focused on the core aspects of character (e.g., self-value, courage, moral excellence, perseverance, discipline, and self-sacrifice), you can turn your attention to improving some general tools.

Expand Your Knowledge Base

The scientific focus of our careers narrows as we devote time and energy to solve particular scientific problems. This is, of course, unavoidable as each scientific field discovers new information that its workers must master.

Its depth increases. However, subspecialization isn't adequate justification for knowing only of the advances in your own field. Tight focus is fine, but tight focus all the time produces limited perspectives.

After all, even though we believe our work is critical (why would we do it if we didn't?), the fact is that by and large, the rest of the world does not understand it and is getting along just fine without this comprehension.

Why not take the time to learn about another area? There are many activities in other fields that are worthy of your understanding and careful consideration.

Developing a different knowledge base offers you several advantages. Your examination of another province provides a useful and timely distraction. By applying the analytic abilities that you commonly use in your own area of expertise to understand the developments in another area, you're actually taking the time to sharpen your skills.

The fact that you have to regain your focus when you return to your own work requires effort, but with that effort commonly comes a slightly different and helpful perspective on the issue in your own field.

Law, art, history, and music are just several examples of areas waiting for your reconnaissance. Hardworking specialists in these fields develop good products, commonly writing for people like you who wish to familiarize themselves with what these other areas have to offer.

A serious examination of this work stretches your outlook, broadens your appreciation, and can commonly help you to find the perspective extension that you sometimes need to have in order to understand a point of view other than your own.

Helping a colleague transforms your spirit of generosity into genuine appreciation and new productivity.

When you're stuck on a problem, the solution may not be to work harder on it, but to instead take your mind off it. Speak to some of your colleagues and peers about what they are doing. You might be quite surprised to find that they have a problem that you can help them with.

Spending 1–2 hours with a peer will in all likelihood produce no real setbacks for your work, but can produce a tangible benefit for your colleague, both in the measurable advancement of their own work and also the immeasurable impact that your unfettered generosity has had.

Managing Zoom and Travel

Of course, the Zoom* experience in combination with the need to reduce travel has essentially warped the business travel calendar and experience. When handled correctly, this can be a welcome change to an itinerary.

Clearly, Zoom decreases the need for travel. Travel time is not just the time to fly. It is, of course, the time to pack, transit to the airport, hours at the airport, and getting to the hotel. This in general can more than double airline time.

---

* Zoom is popular but is not the only cheap, easily available videoconferencing unit around. I use it as an icon for all such products.

Plus, there is the time required for you to prepare your colleagues for your absence, making sure that they have all that they need from you before you travel.

In addition, your family also needs attention in preparation for your leaving.

Zoom reduces this time investment. The Zoom time investment is commonly not much different than that of a phone. And you don't end it with the same fatigue that you used to feel coming home from a meeting.

Thus, from this perspective, Zoom is a great advantage to your time management.

However, technology has nefarious effects. With meetings now taking only an hour rather than days for face-to-face interactions, the temptation is to increase the number of Zoom calls.

This is something you should resist.

For those of you who were working when email was first introduced in the 1990s, email was a simple and useful tool. Rather than make a phone call or write and send/fax a letter that took minutes, you could save some time to send an email. An email generates a short response without the frustrating telephone tag. This was fine as long as the overall communication load did not increase. But email has expanded beyond simple phone call/letter substitution, demanding its own imperative of responses.

We are surrounded with the detritus of this explosion. Inboxes are choked. It is not uncommon to have hundreds of legitimate messages in the morning. Much of this communication is defensive and unnecessary. If the senders have been asked, "Would you be willing to write a letter to communicate this information?" the answer is most likely no.[*]

Armed with this experience, we need to consider how we use Zoom calls. Zoom calls in business are wonderful face-to-face meeting replacements, as well as an alternative to conference calls. Nothing more.

Just because we can't see the specific danger doesn't mean we shouldn't be especially vigilant for the implications.

Air Travel and Family

Zoom does not replace all airline travel. There are some meetings that you will still need to attend in person. The period just prior to an airline flight is particularly hectic. Since you'll be gone for several days, you have to work doubly hard to ensure that (1) you're as prepared for the meeting

---

[*] This is to say nothing about the hundreds of billions of emails sent each day that are slattern, exploitive junk.

as possible and (2) you haven't missed some critical issues at the office that must be addressed before you leave.

Travel is high stress. Unpredictably bad traffic, huge anonymous airline terminals, long lines ending with increasingly penetrant security interrogations, gate confusion confounded with last-minute gate changes, delayed flights, and canceled flights mitigate against us. Infuse COVID-19 concerns, and we see that airline travel (long ago one of the most luxurious ways to travel)* is one of the most stressful experiences that we as a culture inflict on ourselves.

Meetings themselves can require your attention for long periods of time. Of course, during the meeting, you're working to keep up with the events back at your office by email and telephone.** Upon the meeting's conclusion, you can have more meeting-generating work to do when you return. Finally, upon your arrival at home, despite your best effort to stay caught up on the road, there are urgent issues to which you must attend. Make sure one of them is family.

Although its complex and personal and culture-driven nature confounds any real attempt to provide specific guidance, we can nevertheless make a general observation about family. Regardless of continent, country, culture, creed, or religion, the importance of family is central. Avoid the trap of letting dedication to your family be replaced by dedication to your career.

Commit important effort to your career, but devote your life to your family.

Families are demanding; strong family relationships require your continued time commitment, and you, no doubt, already know that you'll need to balance responsibilities at home with the demands of work and career.

Let your default position be that you'll resolve these conflicts in favor of your family, using this competition for your time as an opportunity to reassert the central role of your family in your life.

Consider the following example. An area I have struggled with is the separation produced by my travel. The night before I leave on a trip lasting several days has become an especially close time for my family. We are sure to have a good meal together and spend the rest of the evening visiting.

---

* In the 1970s, flying was considered a privilege. Many used to dress up to fly.

** It is not uncommon now for what used to be called "bathroom breaks" be now called "email breaks."

The difficulty that this presents for my work is that I commonly have unfinished job obligations that must be addressed in order to complete the final preparations for my trip. Try as I might, I can never complete this work during the day, and so I bring it home. Thus, my genuine desire to finish my work collides head-on with my heartfelt need to enjoy my family.

I have struggled with this over the years, trying different ways to resolve this and have chosen to spend the time with them. However, the next morning, I will wake up early, say, at 3:30 AM to complete the unfinished work for the trip. I regain the rest I lost by sleeping on the plane later that day. It took me a long time to devise this plan, but this appears to work best for us.

In addition, time with family should be time away from work.

Working at Home

Not all scientists have the opportunity to work from home. Their responsibilities include intense physical chemistry experiments or the careful observation of microbiologic, virologic, or other research that requires time in the lab and therefore away from home.

Nevertheless, our work environment has been upended by COVID-19. Fear of contagion requires us to move in the direction of isolation. And the availability of the smartphones, high-speed internet connections, teleconferencing, and Zoom calls permits us to work effectively from home. If most of your work involves computing, document generation, and/or electronic communication, society has now provided the tools that we required to carry this out.

However, our experience with electronics provides a monitory for us.

Recall that in an earlier chapter, we observed that the ability to work anywhere can morph into the requirement that you must work everywhere and that working anytime becomes working all the time. An undisciplined home-work environment is rife for this time of work overgrowth, requiring your discipline to work out a work lifestyle.

Critical to working home effectively is to set time boundaries and space boundaries. Be stern in the allocation of time. Set a schedule that has you working the required number of hours, but provides for you the time to be attentive to your family needs. Similarly, if you can, set up a space, separate and apart from your family's foot traffic, that permits you to concentrate on your work to the exclusion of distractions.

However, equally important is to let your family know when you will be done. When you leave home for work, your family knows when to expect you. Working at home without such an expectation builds on the impression that you will be working all day with no expectation that you will have time for them. This makes it difficult for them to give you the space you need.

Also, let your family work with you to plan your schedule. Let them know the time you have to be available for calls/Zoom and then give them free rein with your calendar. You may be surprised. They may be fine with you working after dinner on two of three nights a week if you can free up some morning or afternoon time for them. Your work schedule becomes a joint enterprise that your family supports because they were part of the creation process.

This joint approach can be applied in other settings as well. If your children resist some upcoming travel, then let them participate (Which suit should I take? Pick out a tie for me. I can rent one of two cars, which do you think is best? Which of these flights do you think I should take home?). Not only can this be wildly popular with them but it may spark their genuine interest in your work.

The problem with travel is the disconnect between your children and your family. Their interaction on these travel-related trips allows them to sustain their connection with you. When you speak with them, they will ask, "What do people think of the shoes I selected for you?" There can be animated discussion at home when you return, not about your work, but about their role in it. This is a connection worth sustaining.

I ran into a serious problem with family connection during the 9/11 attacks. I was in Washington, DC, for an FDA meeting when the attacks on New York City and Washington commenced. I was safe in Bethesda, had rented a car, and decided to drive to Houston. My wife and daughter were very upset about the attack and about my safety. Fear opens doors in which the rational mind has little impact.

So rather than have them agonize over each passing in which I am presumably driving through potential war zones, I asked them to help me.

I told them to plot my return itinerary with stops for meals and hotels.

They threw themselves into this task, asking how fast was I traveling. What routes was I taking? Why was

I traveling south and not west? Why not go through Charlotte?

After I answered these questions, they constructed an itinerary for stops, meals, and overnight stays for me that they read to me over the phone the afternoon of the first day of my trip. I followed these instructions, and we talked about them at night.* During the trip, they were much relaxed and expectant of my progress. They have planned the details, and therefore they were no longer loving bystanders, but active members of the journey.

The time passed quickly.

Time away from work means not just being physically but electronically separated from your work activities; while away, skip reading your business email, don't work on reports or manuscripts, and avoid "taking advantage of downtime" to work. Downtime, in order to be effective and provide the break that you need, needs to "be down" and "stay down" and not let you "catch up" while you're away.** New insight resulting in restful and unhurried extended weekends is often better than the pedestrian-paced advances that you might have made if you continued to work through a fatigue-hazed Saturday and Sunday.

Finally, remember that while working hard is an enviable endeavor, being consumed by work has tragic consequences.

> **Devote important time to your career, but commit your life to your family.**

---

* They actually routed me through Dallas, three hundred miles north of Houston to stop and see other family members and to pick up some items for them to bring home!

** As an experiment on a recent summer vacation, I changed my approach to e-mail management. Typically, I regularly read my email while on vacation. This time, not only did I shun email while I was away from office for two weeks, but upon my return, I deleted the seven hundred messages—without reading them! Instead, I spoke with my boss and several colleagues to catch up. I found the vacation uninterrupted, my return to work less stressful, reconnected with my team members, and experienced no measurable loss from ignoring the email flotsam.

12

Collegiality

Collegiality is the state of an open, honest relationship with your coworkers that permits free exchange of information, thoughts, observations, humor, and reflections about the work you all are doing.

Collegiality is not just information exchange—computers do that. Collegiality is not just saying hello each day you see each a coworker. Many strangers do this. Collegiality is a carefully and patiently constructed bedrock that we jointly agreed to construct and that governs our interactions.

Like any other foundation, this bedrock requires attention and sometimes work. It should be strong enough to permit learning to work with a colleague with whom you may not agree and frankly do not like. Getting along with coworkers you enjoy is easy and natural. However, our fellow scientists can be strongly opinionated and prickly. It is all too easy to get deflected from their good intentions by an objectionable personality. Your character is the anchor that keeps you from drifting.

"Educate-able"

I have a bad habit that I just can't seem to break. Whenever I download new music tracks from my favorite artists, almost invariably, I don't enjoy the selection as much as I thought that I would. They are a disappointment, and I wonder what happened to the artist that no longer satisfied me.

However, as I push those feelings away and give the music a chance, I come to enjoy it. In fact, after one or two weeks, the new selections become among my favorites. It's not that I like the artist's older music less; I now simply like them both. I had come to think of this phenomenon as "the music growing on me."

Nothing could be farther from the truth. It didn't grow on me—I grew on it.

The problem was that I could not appreciate the new music with a set of fixed and old expectations. By overturning the resistance in my own mind, I remade myself in this one area.

Scientists to whom I first reacted badly have over the years become my valued friends. These scientists are still as different from me now as they were when I first met them, but they have become my friends anyway. They have grown on me, and I have grown on them.

At this point in your career, you have completed over sixteen years in school (and for many of you, more than that). You're highly trained and have earned the chance to have an impact on society.

However, despite the different training and talents that scientists have, there is one trait that is necessary for you to maintain—you must always be "educable." This is not research "educability" because we each already know that we will as scientists always be learning new science and new technology.

Developing collegial relationships is based on a different style of educability.

When I was in graduate school, there was a man who seemed to me to roam the library. If there were twenty of us students studying separately and quietly in the hall, he would come up to each one and simply ask, "What are you doing?"

I reacted badly to this. I thought he was being nosy, or looking for a distraction from his own work, or looking for a leg up by pirating someone else's. It finally hit me as he did this each day that he was simply trying to be friendly, to understand what his new colleagues thought was important. All that he was doing was trying to get to know us.

I have found that many scientists stay to themselves. At many large meetings, they listen quietly then walk quietly to the next meeting. Yet their nature may not be that way.

They don't want to talk to strangers, but they also don't like being alone.

However, when someone approaches them, engaging them in a conversation that is wholly innocent and nonthreatening, they respond warmly and positively.

If you want your fellow scientists to talk to you, be willing to talk to them. If you want friends, then be willing to go out of your way to be a friend to someone.

I have never known anyone who did not want a good friend. And in these days of additional confinement and COVID-19 concerns, when fears run high, the need for good friends, for connections to people, is strong.

The training of these new colleagues is different, their command of your language is different, maybe their customs are different. They are an unknown to you, and you must replace your vacuum of intuition and understanding about them with new knowledge that you never anticipated you would have to gain. Their perspective can affect your career, and perhaps, more importantly, your interactions with them can change each of your lives if you give them a chance.

As scientists, our only saving grace is that we are "educable."

Idiot Savants

The definition of an idiot savant is an individual who has a mental or learning disability but is extremely gifted in a particular way, such as performing feats of memory or calculation.

Neither you nor I are idiot savants.

Having said that, we must acknowledge that we spend much of our intellectual development in one field. For example, I have spent the past three years immersed in something called "the duality principle." Therefore, I have not spent time in commerce law, the theory of classical music, Soviet military doctrine of the 1980s, or the routes de Soto took in his exploration of the new world.

We are not idiot savants, but we have to acknowledge that our knowledge depth is deep in a small number of areas, but shallow in most others.

Yet the lay public does not treat us this way. They tend to believe the reverse; our superior expertise in one area actually can be assumed to apply to other areas as well. So we are commonly asked cultural questions, sociologic questions, technological questions, or even political questions, areas in which we are not experts.

Don't fall into the trap of inappropriately broadcasting expertise that you don't have. Make sure that you understand what you know and what you don't know. The audience may think you have all the answers, but make sure you understand your own limits.

There is never any harm from honestly answering "I don't know."

Be a Colleague to Make a Colleague

We all want the support of good colleagues. We want to profit from their wise advice, receive their help when we ask for it, rely on their support when our own actions do not rescue us from professional problems. Ultimately, we would like to be enmeshed in a web of responsive collegiality on whose support and strength we can draw. However, in order to be surrounded by this web, we must be part of the web. Just as we desire collegial support, we have to recognize that we must provide support for other scientists.

Following the adage "to make a colleague, you must be a colleague" produces a wonderfully rich working environment. In fact, you can be the spark that helps to start and generate this environment. However, be forewarned that this requires openness, a spirit of generosity, and unselfishness as you bend over backward to help others with whom you work. Genuine empathy coupled with your thoughtful consideration as a fellow scientist to help a colleague through a difficult period can take time from your own work. Your own productivity may falter for a few days.

Let your character development support you here. It is both instructive and satisfying to see that this temporary reduction in your productivity does you no lasting harm because the diligent research efforts that you apply at all other times will sustain and support the temporary decrease in your productivity now. Remember that your long-term goal is not isolated productivity. Since productivity is not the only essential ingredient of professionalism, it must occasionally be set aside to allow other core constituents of your character to appear and exert their influences.

One of these components is the compassion that you draw on to help a colleague. The more strained the atmosphere, the greater the difference a compassionate act makes.

> **Since productivity is not the only essential ingredient of professionalism, it must occasionally be set aside to allow other core constituents of your character to appear and exert their influences.**

Visiting junior colleagues is particularly instructive. They have a unique burden and a stupendous learning curve to climb. Plus, absent any experience in career advancement, they can easily suffer from a lack of self-confidence.

Having a colleague come by with no agenda other than to see how they are doing and offer some assistance provides a "multidimensional boost." First is the tactical support if you can help a colleague work through the sometimes-tortured process of getting a new university computer or laboratory part, or decipher a complex budget, or explain what indirect costs are can provide them intense relief while saving them important time.

However, just as important is how you do this. You were not compelled to meet with them, and they know that. Your voluntary sacrifice of your time is like a mainline "value-injection" for them. Starved by the absence of feedback, disappointed by a manuscript's negative review, junior faculty can be riven by self-doubt. A spontaneous encouraging visit can have a lasting effect.

Another effort that I have found junior faculty appreciate is a colleague reading a paper of theirs that will soon be submitted. Not only is the input of a more published faculty member helpful, but the time taken to sit down face-to-face with the junior faculty member and talk to about the approach to writing and to responding to reviewers can endure.

Gentle Communicator

Emotional considerations count in our interpersonal communications. While we are scientists, at our core, we are simply people who develop and convey scientific information.

The fact that we are scientists does not free us of the responsibility of treating each other with respect and decency. Since much of the scientific information we convey to each other is emotion-free, we must take greater care in delivering the message so that our intent is not misunderstood.

Keep in mind that what you mean to transmit when speaking may not be what is received by the listener. This is a common source of misunderstanding. Your own review of the degree to which you convey your intent should be an important component of your self-assessment.

For example, I have learned that I tend to underconvey the emotional impression that I hope to leave with other scientists. When I am attempting to be courteous, I do not come across as being courteous and kind; instead, I come across as simply being "OK." Even though in my own mind, I believe I'm being courteous, that isn't the countenance that is perceived by others. On the other hand, if I speak or carry myself in a manner that leads me to believe that I'm being "overly courteous," then I am seen not as being obsequious, but instead as being pleasantly courteous and friendly. I don't quite understand this, but having recognized this about myself, I take this into account in my discussions.

Apologies

Conversations are so easily facilitated by a simple and honest apology for any misunderstanding. We all know the fortunately rare and rarely fortunate scientist who believes that standing on principle gives him or her the right to be rude and offensive.

Whenever it enters your mind to apologize, act on that thought rather than dismiss it.

A sincere apology goes a long way to smooth a rough conversation. It doesn't mean that you have to give way on a matter of conviction. Apologizing simply reveals that you have the strength to recognize that you might have caused offense while standing your ground. It also announces your intent to repair a situation or relationship. And your character development reveals that offering a sincere apology does not reduce your value.

Consider apologizing when you believe that you have done nothing wrong and therefore an apology is not warranted. For example, you may have merely attempted to criticize a small technical detail of an otherwise fine report; however, the authors perceive your words as a personal criticism of them. This is not your fault, and you did not mean to convey any insult, but they nevertheless take it personally.

This is very easy to do if you have a solid sense of self-worth that is independent of the point of view of others. With this stanchion, you can easily apologize because you are not threatened by suggesting if not wholly admitting that you made a mistake. Investment in your own character permits this.

If you abhor mistakes, then I fear that you are in the wrong field. Science is rife with mistakes. You will make many.

Work on developing an intuition of how others perceive you, striving to adjust your speech and tone so that what you communicate is what you intend to convey. Purge any bellicosity from your speaking patterns.*

Having worked to ensure that you have the best, honest, and benevolent intention in speaking, ensure that the same well-meaning intent is clearly conveyed in an easily understood way. If it is not, then be quick to apologize for the misunderstanding and try again.

If I can't speak this way, then it's best if I keep my teeth together and say nothing.

Manage Your Anger

During your development as a scientist, you'll become angry, and you'll arouse anger in others. This anger is readily apparent in discussions at professional meetings, in which the sharp exchanges between scientists can be tinged with bitterness. Rarely is the motivation for the anger clearly in view, but its presence is undeniably palpable.

When anger is present, it gets in your way. Scientific exchanges are never enhanced by anger. It is more common that the scientist who is attempting to get his point across through his anger has more difficulty enunciating the point clearly.

Also, sadly, the scientific point she is making is commonly overshadowed by the reaction of her fellow scientists to the anger. Most of the conversation after the exchange is about the anger and not about the illuminating points made in the debate.

Anger can also arise in a group of scientists working under tremendous time pressure. With a deadline looming or an additional workload added, there may not be enough time for you to accomplish the desired task. Frustration and anger can quickly rise up in this circumstance. This anger can ruin your relationships with your colleagues.

Before we discuss what to do about anger, let's first distinguish anger from the energy that comes from a healthy debate or a challenge. Both can excite you. However, the productive energy that is generated in you from facing a challenge animates, spurring you on to a new line of thought, driving you to reevaluate your own scientific theories objectively.

---

* This has been discussed earlier.

On the other hand, anger taps into the belligerent ideas of attack and defense. Facing a challenge helps to polish your ideas, theories, and concepts. Anger, like lye, dissolves them. A good challenge produces productive energy. Anger produces a self-destructive force.

What can you as a scientist do to manage your anger? One way to accomplish this is to predict it. The simple recognition that a situation in which you'll be involved may produce anger can be defense enough against its occurrence. Anger commonly presents itself as a reaction to a surprise. If the surprise can be anticipated, then we can block the angry reaction.

Secondly, don't be compelled to respond to a harsh word angrily. Those scientists who have the reputation of losing their self-control, lashing out angrily in response to criticism, essentially give over their self-control to others. Their critics can easily control them by speaking softly or harshly, depending on their desire. Responding in kind does nothing but bring out the worst in you and diminish you in the eyes of others who are observing the exchange. And your character development revealing that anger toward you does not reduce your value. With solid self-value, what do you lose in public criticism directed to you?

As a scientist, the best reaction to harsh verbal and public criticism directed to you is a kind word. A humorous reaction, uttered kindly, is another healthy response. While this caliber of reply disarms and disorients your critics, elevating you in the eyes of observant others, it is also a reply that is good for you. By resisting your adversary's attempt to plant a seed of self-destructive anger in you, you have looked after and protected your own well-being in a difficult setting.

Finally, if these tactics do not work, then recognize that the anger you feel is getting in your way and release it. If exercise works, then exercise.[*] Try to do nothing constructive until you have purged yourself of this difficult emotion. Like poison, it is best to expel it as rapidly as possible.

Conclusions

Collegiality is central in professional development. Deliberately choosing to turn your face away from your own productivity (and your own needs) in order to come to the aid of a colleague not only helps a coworker who is in some distress, but is also good for you. Setting aside the fact that

---

[*] For me, it's a long drive or a hard workout on a punching bag (which, fortunately, never punches back). I feel terrible when I start and relieved when finished.

your stature increases as a result of your generous action, placing your own needs voluntarily aside, produces its own character reward.

Focusing on developing the strengths, skills, and outlook of a mature, professional scientist will not only amplify your consistent productive efforts, but will also buffer and protect you as you face the unseen challenges that lie ahead. The specific scientific advances that your work produces will, in all likelihood, be overshadowed and surpassed by the future advances of others. However, the principles for which you stand as both a scientist and as an individual can resonate indefinitely.

13

Investigator as Collaborator

Congratulations. You have been invited to join a research group.

Whether an investigational team is newly constituted or has a well-established productivity track record, the invitation for you to become a fellow member is gratifying. This chapter focuses on the particular pressures that you as a junior scientist face in these collaborative environments and, more importantly, how you might prosper through your involvement.

To a very great degree, your success depends on your sense of self-worth and value that you have discovered through your character sabbaticals and development.

Character growth requires that you develop the stature to respond with stability in a tempestuous intellectual environment. The ballast of an independent sense of self-worth helps you to maintain your balance in the face of these unpredictable interactions.

Plunge In

Once you have developed and strengthened your source of self-worth, move ahead into your new role as a collaborative researcher.

If you have several competing opportunities for collaborative effort, inspect each of these closely and carefully. Take the time to have several detailed conversations with investigators on each of the projects. For each potential project, more than one conversation will in all likelihood be necessary. The first conversation will require you to listen to a broad description of the research effort.

This presentation by investigators who are already involved in the study will cover the purpose of the research, the contribution that the project is expected to make to the scientific community, and the opportunity that the research will provide for your own productivity and advancement. Your solicited involvement might be portrayed as a "win-win" scenario; you win by having the opportunity to increase your productivity, and the project wins by having you as a team member.

While it is not quite fair to call this conversation your investigator has with you a "sales pitch," the conversation nevertheless has the definable features of a promotional talk.

In this preliminary conversation, the research program is often presented in its most favorable perspective. This is not to be unexpected since the experienced researcher strives to attract you to their project.

However, you should not dismiss the discussion simply because the project is presented in its most auspicious light; in fact, you'll gain a good education about the scope of the project from this early conversation. However, keep in mind that this discussion is not an end, but is instead a preamble to the more important conversations that must follow.

The subsequent conversations about the project are critically important because it is in these conversations that you learn the role you'll play in the project. Specifically, the central topic of these conversations should shift from "What is the research program about?" to "Exactly what will be asked of me in this project?"

Be persistent in eliciting exactly what you'll be doing in this project, to whom you'll report, and a general description of the important timelines of the project. An important goal of this conversation should be an understanding of the percent time and effort that you should exert on this project. In your conversation with the researchers, do your best to pin them down. This critical information is central to your decision to participate in the project.

Junior investigators are understandably sensitive to irritating their prospective new senior researchers with these types of questions, but clear and early responses to these interrogatives can avoid serious and unfortunate misunderstandings. A useful way for you to begin this conversation would be as follows:

> Investigators in other projects have disappointed both their principal investigators and themselves by making inappropriate assumptions about the time commitment to this project. I would like to avoid making that mistake, and therefore wish to know what is the time commitment that you expect of me?

You need a frank, realistic, and cordial answer to this important question; and many times, this is exactly what you'll receive.

However, unfortunately, there are some senior investigators who do not react well to these queries. This latter group of investigators may believe that such questions are "piker" questions; that is, that they betray a sense that the new investigator is interested in working on activities outside of the proposed research project at hand, that they are not fully committed to this research.

Treat this type of reaction as a "danger" sign. These senior investigators often labor under the false assumption that junior investigators should be willing to provide their complete and wholehearted support for the senior investigator's project in order to gain the experience and training that comes from this involvement.

From this perspective, the junior investigator should be at their seniors' beck and call, expending many hours, evenings, and weekends on the research project, working above and beyond the degree to which the junior scientist is paid.[*] According to the senior investigator, all this time and effort is justified by the new knowledge and experience the junior investigator obtains.[**]

---

[*] Think of the junior example in the first scenario presented in this book.

[**] A fine example of this philosophy is in medicine. When a young surgeon in training dared to complain to the senior physician that the process of staying in the hospital every other night was reducing the young doctor's educational experience, the older physician exclaimed, "We are doing you a disservice by letting you leave the hospital every other night because by doing so, you're missing half of the patient admissions!"

A difficulty with this approach is that the junior scientist is not compensated in accordance with their work. This apprentice-style training is still prevalent in and perhaps can be justified in a student's education. However, when the training is complete, the junior scientist should receive adequate compensation for their time commitment.

The propensity for some senior scientists to treat junior researchers like students can be the source of difficulty.

The best way that you as a junior scientist can deal with this problem is to face it early and openly. Take an honest, collegial, and direct approach in exploring this issue with your senior investigator, explaining your point of view to these senior scientists. Articulate that you have a solid commitment to work on his project, a commitment that you plan to keep. However, you also have a commitment to work on other activities, commitments that you must keep as well. Explain to her that you take each of your commitments seriously and therefore must ensure that you have adequate time to fulfill them all.

Be lucid about your expectations from the project and understand clearly what your principal investigators anticipate for your workload.

Junior investigators can balk at this recommendation. Some believe that they don't have the courage to push the conversation through to the end; others believe it is inappropriate to discuss the matter of time commitment so candidly with a senior investigator.

However, courage can commonly be bolstered in the realization that the alternative to a direct approach is far worse in the long run.

With no such conversation and understanding, you may find that you're required to work increasingly long hours on this project. In the absence of the conversation that would have set the boundaries of your work, the investigator does not understand your need and fills this void with his own expectations. He mistakenly believes that your commitment to his project is open-ended.

This assumption on his part is to your detriment because it places increasing demands on your time that you cannot meet. Some of the work that you have been asked to do you cannot accomplish because of time commitments, leaving some of the work the principal investigator asks you to accomplish undone.

Ultimately, there might be a frank, candid, and blunt conversation between the two of you. As the exasperated junior scientist, you require relief from the overwhelming workload.

Unfortunately, your request falls onto the ears of an equally exasperated senior scientist who wants to know why you're unwilling to keep your commitment. This unpleasant circumstance is made all the worse by the recognition that the entire affair could have been avoided by a simpler conversation much earlier in your relationship with the team leader.

It is better to have a challenging conversation early in the project than a rancid one late in your experience in the program.

If your best effort to convey both your sincere commitment to work on his project and your need for an understanding of your limited-time availability is greeted with scorn or derision by the project's senior investigator, and you do not have unlimited time to work on the project, then leave. Work on another project that allows you to fulfill your time commitments.

Better to separate while you're still on speaking terms than to let your relationship degenerate to an angry and hostile one.

It is possible that you'll have several competing opportunities from which you must choose one program in which to work. While a component of your decision process will be made up of considerations of what would be best for you, it is worthwhile to avoid having this style of thought dominate your decision process.

Refamiliarize yourself with the goals of your department, as well as the goals and mission statements of your institution, seeking an answer to the question "What is the best project for my organization?"

Finally, avoid the deliberation suggesting that your future depends on your choice of product team or research group with which you choose to work. While the program you choose may have some influence on your development, its ability to set your career trajectory is exceeded by the influence that you yourself exert.

It is the combination of your abilities, your capacity to learn, your communication skills, and your intuition that have the dominant role in determining your career trajectory, not the particular project on which you're working. Your chosen project's effect on your career is much like the influence of the wind on a high-performance jet aircraft as it flies to its destination. The wind may slightly increase or slightly decrease the duration of the flight, but it has no important impact on the plane's arrival at its destination.

In fact, at this state of your character development, you may be better served by a position on an obscure little-noticed, but useful project than on a project that has high visibility but an overbearing production schedule. Choose a project that requires you to work at a level that is greater than your current level of operation but within your reach.[*]

Begin to gain a sense of boldness that is coupled with a grasp of the achievable.

Once you have chosen your project, immerse yourself in it. Plunge in. There are many ways to learn about swimming but only one way to swim, and that involves getting wet. Like the new (and sometimes not-so-new) swimmer, you'll occasionally swallow some water. That's an expected by-product of the learning process, so don't be deterred by the occasional difficulty. Cough it up, spit it out, and splash ahead anyway.

> **While the program you choose may have some influence on your development, its ability to set your career trajectory is exceeded by the influence that you yourself exert. It is the combination of your abilities, your capacity to learn, your communication skills, and your intuition that have the dominant role in determining your career trajectory, not the particular project on which you're working.**

Mastery

The first important step that you can take in your new project is education. Educate yourself with the science of your new collaborative research effort.

Since none of us knows everything, ignorance is no vice. However, being comfortable with ignorance is.

This education process begins with the acknowledgment of what you do not know. This can be a sensitive issue for you because as the junior investigator, you may already feel insecure about your new appearance before investigators who are perhaps all senior to you and have already established cordial relationships with each other.

However, if you have worked hard and patiently to develop your self-esteem and confidence as outlined earlier, the new obstacle presented by

---

[*] A later chapter discusses some notorious examples of junior scientists who were caught up in projects too demanding for them and the unfortunate consequences.

your lack of knowledge is only a temporary impediment to your progress. Since your source of self-worth is from within yourself and not from without, you can quickly gain the knowledge that you need without being crippled by insecurity.

How you conduct yourself during this phase can be critical to the establishment of your position in the research team. As you learn, be sure to admit your missteps easily and readily. Senior scientists recognize the value of a colleague who easily admits mistakes and is self-correcting.

> **Senior scientists recognize the value of a colleague who easily admits mistakes and is self-correcting.**

Learn at Your Level

Before your first meeting with your new coinvestigators as a new member of the team, you'll naturally want to spend some time reviewing the relevant literature about the new topic that is the subject of this research effort. If the topical content is technical and complicated, then there are some important steps that you can take to become acquainted with the new knowledge base.

First, find a book or other source that is at your level and that you can understand without too much difficulty. One of the most frustrating intellectual experiences for anyone is to attempt to learn material that is presented at a level that is just too advanced and difficult for them to understand.

My most disappointing experience in learning a scientific topic at any point in my career was my first effort to learn measure-theoretic probability. This advanced topic required not just a probability prerequisite, but a formal mathematical analysis background as well. I lacked this latter background and therefore was unable to satisfactorily progress through the subject matter. I'd spend three hours in intense study, and at the end of the study period, I gained only three minutes' worth of new and useful knowledge to show for the effort.

My intense desire to learn the material, coupled with my substantial time commitment, could not overcome the shortcomings of my poor background. However, once I recognized that I was missing the necessary mathematical prerequisite, I brought my fruitless efforts to learn this complicated topic to an end.

Holding my readings in advanced probability in abeyance, I took the time to identify an analysis text at my level. Once I had covered and absorbed this material, I was able to return to the measure theoretic material, this time making the consistent and steady progress that was my initial goal. More importantly, I had rediscovered my self-confidence during this relearning process. Acquainting yourself with hard material is, of course, much easier if you start with material that is within your reach.

Stretch yourself, but not to the breaking point.

> **Stretch yourself, but not to the breaking point.**

Systematic Reviews of Manuscripts

It can be very useful to carry out a systematic review of the efforts that have led to the research project that you have recently joined. This comprehensive review can involve reading many different manuscripts that cover the important information and describe the relevant advances in the research area. Your review of these manuscripts must be efficient and brisk while simultaneously engaging your abilities of discernment and integration.

There are many ways to engage in a systematic review. They involve both readings and discussions. I would recommend that you begin the process by first reading the literature, well before you start a dialogue with others in the field. Identify a universe of literature proceeding in chronologic order. This is how your more senior colleagues, who have spent a good deal of their time in the field (most likely contributing to the literature base that you're studying), have developed their knowledge. This methodological process also makes it less likely that you'll have missed a key step, a mistake that is more likely to happen if you leap right away to summary or review articles.

Your purpose in reading each manuscript is to identify the contribution the research effort has made. Specifically, this general evaluation reduces to three questions that should be answered:

1. What was the purpose of the research?
2. Do the methods used allow the researcher to answer the question?
3. If so, then what is the answer?

The motivation for the research is determined from a careful study of the introductory section of the manuscript. Commonly replete with citations, this section clearly delineates the research question raised by the investigators.

Whether the investigators are able to answer the research question they asked is determined in the methods section. Your review of this section of the manuscript should be careful and thorough. Was the instrumentation sufficient to provide the measures with the precision that the authors require? If laboratory samples are involved, were they preserved using the appropriate environmental conditions?

If the research involves subjects, other questions must be adequately addressed in this critical section of the manuscript that describes the methodology used by the authors. Was there an adequate number of subjects? Is the analysis carried out prospectively delineated, leading to trustworthy estimates of effect sizes, standard errors, confidence intervals, and *p*-values; or were the analyses exploratory, requiring an additional confirmatory study to sustain the findings?

If the methodology empowers the investigators to answer their prospectively asked question, then proceed to read the results section, fully planning to integrate the main findings of the research into your fund of knowledge. However, if the methods reveal that the findings from the study are exploratory and not generalizable, then save yourself some time and set the manuscript aside. Findings that are intriguing but cannot be extended to the larger population have no real place in your systematic review.

Avoid the temptation to "study count." "Study counting" is the process of simply counting the number of studies that address an issue and deciding if there "are enough" studies to support the result of interest. Some who are involved in study counting argue that there must be more than one study.

Others say that the number of "positive" studies must outnumber the number of "negative" studies. Instead, the scientific reasoning process assesses in detail each of the available studies, carefully dissecting the methodology, sifting through the results, and carefully considering the conclusions.

Study counting represents the wholesale abandonment of the intellectual principles of careful review.

The specific problem with study counting is the implicit *ceteris paribus* (all other things being equal) assumption; that is, that all the studies that are being included in the count are equal in methodology, equal in the thoroughness of their design, equal in the rigor of their execution, and equal in the discipline of their analyses and interpretation. This fallacious assumption is far from the truth of scientific discovery. Studies have different strengths and different weaknesses. Different investigators with their own nonuniform standards of discipline execute the research efforts. Some studies give precise results while others are rife with imprecision. The panoply of studies is known not for its homogeneity, but for the heterogeneity of designs and interpretations.

We must distinguish between the appearance of an isolated study; that is, one study whose finding was contrary to a large body of knowledge available in the literature and the occurrence of a sole study. There is ample evidence that a sole study, when well designed and well executed, can be definitive [1]. What determines the robustness of a research conclusion is not the number of studies but the strength and standing of the available studies, however many there are. Science, like the judicial system, does not merely count evidence—it weighs evidence. This is a critical and germane distinction. Study counting turns a blind eye to study heterogeneity.

Another useful approach to the critique of a research effort is as follows: Start with a review of the hypothesis and goals of the research effort. Then before proceeding with the actual methodology, turn away from the manuscript and begin to construct for yourself how you would address the hypothesis. Assume in your plan that you have unlimited financial resources, unlimited personnel resources, and unlimited subject resources. Think carefully and compose the best research design; that is, the design that builds the most objective platform from which to view the results.

Only after you have assembled this design yourself should you return to read the methodology, the results, and the actual interpretation of the research effort that was executed. Having constructed your own "virtual" state-of-the-art design, you can easily see the differences between your design and the one that was actually executed by the researchers. After identifying these differences, ask yourself whether and how the compromises made by the researchers limit the findings of the research effort. If they pose no limitations when compared to your design, the researchers did a fine job by your standards. If the researchers have substantial limitations, then the analysis and its ability to be generalized may be crippled.

The researchers may have had understandable reasons for the limitations, but these limitations can nevertheless undermine the strength of their findings.

Avoid studying review or summary manuscripts until the end of your review of the original literature. Review manuscripts are fine, but they represent another writer's distillation of the literature. While you can be informed by that summary, first allow yourself to draw your own conclusions about what the data in the research efforts mean.

Once you have completed your own synthesis and integration, pause and take the opportunity to carefully think through what you have read and develop an opinion about the implications of the literature that you have studied. From each of the manuscripts, intellectually stitch a composite point of view that includes the reliable findings from these studies.

Then and only after you have completed this mental synthesis, read the review manuscripts of others, focusing on the differences between their findings and the independent distillation that you have created. These differences may reveal weaknesses in your own integration of the literature that you have reviewed thus far or represent something that the summary authors either missed or inappropriately discounted.

The "Dumb Question"

One of the single greatest impediments to the junior scientist on a research team is their pertinacious reluctance to ask a question that they think is "dumb." After spending time reviewing the background material, the progress of the junior researcher may be blocked because there is a small intellectual impediment. This block can be relieved by an answer to a "dumb question."

Yet the junior investigator is commonly reluctant to ask the question for fear of embarrassment and ridicule. The inability to ask the question reduces the junior investigator's effectiveness and can lead to even greater embarrassment in the long run. The reluctance to ask "the dumb question" commonly finds its justification in the fear of humiliation and failure, and the presence of this fear reveals a dependency on the opinions of others that should be addressed as a matter of character assessment and growth.

The unwillingness to ask "the dumb question" is an understandable tendency that you as a junior investigator must resolutely overcome. Any impediment to your ability to absorb a new fund of knowledge should be removed at once with the smallest possible hesitation. While I would not suggest that you'll ever enjoy asking questions at this seemingly elementary level of inquiry, neither should you be ashamed.

As before, your source of self-esteem is critical here. You're new to the research material, and you have set the commendable and achievable goal of mastering the background knowledge base. Take control of your emotion, banish the embarrassment, ask the question, and remove the impediment. Don't be like the person who would rather stumble around blindly in a dark room. Rather, ask the simple question, "Where's the light switch?"

Another reason to ask this style of question is that you as a coinvestigator must have all the relevant information that you need in order to make a solid contribution to the product. If you're a subspecialist, you may be required to make a technical contribution to the project. The details of the technology may not be understood by the investigators. Examples of these contributions are mathematics, database and computer technology, biotechnology, and statistics. You must understand important details of their work in order to custom-fit your contribution to the needs of the project.

Any misunderstanding on your part about the research program at the beginning of your participation can lead to a misapplication of your contribution.

This weakness may not be visible toward the beginning of the project, but may be quite apparent at the project's end when it is time to disseminate the project's efforts and results. Whatever the forum of promulgation, be it onsite reviews, presentations, or publication, there will be the necessary and important opportunity for criticism. Your coinvestigators, recognizing that an external dissection of the project is about to commence, will ask themselves and each other important questions about the project to make sure that the program is solid and that all of its pieces fit together to form a cohesive whole.

Thus, at the project's end, you can expect many questions from your senior colleagues as they assure themselves that they understand the nature as well as many of the specifics of your contribution. They must do this as part of their preparation for the questions that will arise from outside discussants about your work. These questions will occur at all levels. Your colleagues will not (and should not) be embarrassed to ask their questions of you. Anticipating their questions, you should not be embarrassed to ask your questions of them now in the beginning of this project so that you're prepared for the intense intraproject scrutiny at the end.

Commonly, as a new member of a research program that is being conducted by senior scientists, I will interrupt a conversation that is conducted in my presence, but whose content I am expected to absorb, with the statement:

> I'm sorry to interrupt here, but I have a couple of dumb questions.

I then proceed to ask these questions. I ask them as plainly and as clearly as I can, ensuring that there is no evidence of frustration or impatience in my voice.

Uniformly, every time that I have done this, there have been the same three responses. The first response is that my question is answered. It is answered without sarcasm, with a voice that instead of dripping with acrimony is gracious and charitable. If I have any other questions, I then ask, "Can you suggest where I can read more about this?" The response to this question has, again, always been generous and informative.

The second response is that someone else who was also present during the conversation, but who remained mute, will quietly say that they are glad I asked what I did because they had the same question. It always turns out that the "dumb" question wasn't so dumb after all!

Lastly, an additional by-product that is produced by asking a low-level question during these important conversations between the more senior investigators in the project is the effect that the question itself has on the conversation. Asking "the dumb question" does not reduce the intellectual level of the discussion, but it loosens the conversation. No one was willing to ask their set of "dumb questions" that were different than mine until I chose to vocalize my set.

Once my questions were out on the table and answered collegially, the other investigators relaxed enough to ask their questions. It is as though the ice has been broken, and others are now free to reveal their ignorance and vulnerabilities to each other without fear of rebuke.

The conversation becomes more collegial, more relaxed, and more importantly, a true educational experience for everyone. Much good product is produced when you courageously ask your simple question. Your choice to ask the question is not just good for you; it's good for the study.

Here's the bottom line: You'll probably be embarrassed to ask the question. Go ahead and be embarrassed, but ask the question anyway.

Feel abashed, but don't let that feeling stop you from gaining the knowledge and understanding that you need to be a full participant in the research project.

Teach in Addition to Being Taught

During this early phase of your work in your new project, you will, in all likelihood, not be the only investigator who is absorbing new material. Since you'll be making a contribution to the program, the other coinvestigators will need to learn of your contribution as well.

Just as the program is stronger with your knowledge of the project, the project will also be stronger with these scientists' understanding of your contribution. You may be able to avoid their critical misunderstandings of your contribution by giving them a lecture, the purpose of which is to provide a brief primer in your area of expertise.

As with many areas of their interactions with these more senior investigators, junior investigators can be easily intimidated by this experience. The idea of providing a brief lecture on your area of expertise to the senior investigators can be intimidating.* However, the skills that you gain from that discussion will help to drain the stress out of this presentation.

At the conclusion of your talk, you'll have accomplished two things. The first is that your discussion will have been well absorbed by an attentive audience. The second is that the senior workers in this research effort will gain a new appreciation of both your courage and your capacity to communicate relevant and perhaps complicated information to a challenging audience.

Reliability

> *Nothing astonishes men so much as common sense and plain dealing.*
>
> —Ralph Emerson

The productivity of your cooperative program is essential, and your work as a junior investigator will contribute to that effort. In fact, your product will have to dovetail with that of others in order for this joint effort to generate a noteworthy result. With each investigator working toward a

---

* A later chapter will focus on how to make presentations to audiences of all sizes.

common result (or set of results) for the joint program, each investigator's work effort must fit together with that of the others.

Given the independent work efforts and ideas of the participating investigators, this collaborative and interactive effort requires careful planning by the project leaders. During this planning effort, you'll be asked to produce a set of deliverables. An important measure of both your maturity as an investigator and your stature in the program will be your reliability.

Do What You Promised

At the bare minimum, reliable investigators do what they said that they would do for the project. They deliver work products on time and in the anticipated format.

Unreliable investigators can bedevil a project. They do not provide the deliverable that they claim they can supply or provide the deliverable too late or in an incorrect format.

The investigator whose contributions are chronically missing, or late, or arrive in an unhelpful configuration commonly denies the impact of her effort's unreliability. She minimizes the impact of his nonproductive activity and claims that she will meet the next deadline. However, with each passing unmet target date, the situation worsens, and the quality of the conversations degrade. Finally, it can be difficult for this unreliable investigator to make a persuasive point about any ongoing aspect of the project as long as the project is waiting for his required part.

The project cannot put its weight and foundation on this unreliable investigator to aid its progress. Like a leg that's asleep, it doesn't bear the required weight at the required time.

There are three keys to being reliable: planning, diligence, and above all, attitude. Of these three, the first two are the easiest to attain. Anticipate that you'll be asked about when you can provide the necessary product for this work effort and give careful consideration to this question. While the discussions of the aggressive timelines of other dynamic investigators in the project can be infective, try to avoid losing yourself in this excitement; this enthusiasm can blind you from a clear view of what a reasonable timeline for your contribution will be.

Only after you have given careful thought to your own various and critical time obligations should you commit to a specified date for your product.

Think carefully before you speak because you'll be held to what you say.

A useful tool implemented among engineers planning complicated projects is building in the time and opportunity to deal with unexpected emergencies. Despite your best efforts, equipment can break. Dogs don't eat homework anymore, but the arrival of reagents can be delayed. Hard drives fail. Computers are stolen. People get sick. These occurrences represent the unpredictable happenstance of life surrounding you.

Who among us would not take traffic patterns into account when we agree to attend an important meeting requiring a drive across town? You may plan on a direct route, but you must also plan for the impact of others on your progress. Build in the time to deal with emergencies and other unpredictable events.

> **Build in the time to deal with emergencies and other unpredictable events.**

There are many electronic devices that are now available that allow you to plan and schedule your work, whether that work spans days, weeks, or months. It doesn't matter which of these electronic assistants that you buy. However, whichever device you choose, it will only help you if you affirmatively and actively decide to use it assiduously. For me, this means making only the notations that matter. I only enter those items into it that I fully intend to accomplish. Making a new entry on my calendar is my commitment to myself to accomplish the task that day. If there is an item that I am not willing to diligently resolve, it doesn't get into my calendar.

To help you stay on track with your production schedule, plan on providing regular and frequent progress reports to the program team. In fact, you may be required to update the team leadership, along with all the other investigators on the team. If you're not, offer to give a terse summary of your work.

The purpose of this volunteer effort is not self-promotion. If you have misunderstood any aspect of your role or configuration of your deliverable, you'll certainly learn of it during these brief reports. Of course, the additional practice of speaking to the group, listening to and absorbing their comments and criticisms, will help to build your growing confidence as well as their confidence in you.

Meat Hooks

Once you have planned carefully, carefully follow your plan. Find and insist on an isolated time to work on the project. Ensure that you have the necessary tools to do the job. Create the environment in which you can bring your combination of knowledge, talent, intuition, patience, and consistent effort to the task at hand. If you have telephone/visual conferences, schedule them so that they do not interrupt your work. For a time, let your phone ring unanswered and let your email pile up. However, you probably can't get away with what one famous scientist did.

> On a day when he was receiving visitors, Albert Einstein allowed a fellow scientist to come into his work parlor for a friendly discussion. When they entered his office, Einstein asked his colleague to please wait for a brief time while he finished up a small matter. As the visitor watched the great scientist work, his eyes were drawn to the features of Einstein's workroom. Prominent among the trappings of the room suspended from the ceiling, was a huge iron meat hook. Stuck onto the hook appeared to be a collection of papers. The papers did not appear well organized at all, each sheet being pierced by the hook in a haphazard fashion.
>
> When Einstein turned to face his visitor, the visitor asked him what the papers hanging on the hook were. Einstein replied, "They are letters and correspondence that I have received, but have not yet had the opportunity to answer". Noting the depth of the pile of papers, the visitor asked Einstein what he did when there were so many letters that they no longer fit on the hook. Einstein mirthfully responded, "I burn them."

While we can't get away with this behavior, there is no lasting harm in shutting down your email client for a time or having your phone messages go to voice mail while you quietly prosecute science.

Remember—working on the science is supposed to be the fun part of the job, the portion that you should enjoy. Devote yourself to the task while simultaneously dedicating yourself to completing it in the allotted time. While you're working, keep a running tab or list of new ideas that you get[*] but cannot act on because of your concentrated effort on the topic at hand.

Finally, enjoy these few hours because as you have already seen, much of your time is spent in necessary, but not necessarily scientific, endeavors. Relish the moments you get to bring the full strength of your scientific talents to bear on science questions.

Above and Beyond

As we have seen, an important goal for you as a member of the team effort is to produce the required contribution from your work effort on time and in the proscribed and expected configuration. However, this is only your first goal. As a maturing investigator, you should also have the goal of helping your colleagues produce their required contributions on time as well. This additional step is above and beyond the call of duty as a junior investigator, but is a worthy goal for a maturing scientist who wishes to be a valued member of the research team.

This second goal requires devoted and consistent effort on your part, working to ensure that your contribution will be ready on schedule. As you planned your productivity schedule, recall that you built in time that would be required to deal with the vicissitudes of life. If you were fortunate enough to have a peaceful and unhurried period of time for you to do your work and you were diligent in your effort, then you may have some time to pitch in and help others who either did not plan carefully enough or were overwhelmed by other events. If so, then step in and help. After all, there will in all likelihood soon be a time when the world catches up with you, and you'll need all the help that you can find.

Be the type of scientist on whom your superiors can rely. If they don't notice this ability of yours, be reassured that others will.

Falling Behind

However, just as there are times when you'll be ahead of schedule, there are, of course, also the times when you'll fall behind. There will be circumstances when your day is too full and falls apart. Work that you had planned doesn't get your attention. As you rise in your field, you'll face this distressing event with increasing frequency. Thus, you need coping mechanisms for falling behind your carefully planned schedule.

---

[*] I have a relatively inexpensive voice recorder allowing me to and store ideas as I get them. Later, I can listen to them, sorting, developing, or discarding them at my leisure.

A clear and useful way to deal with this chronically occurring source of dissatisfaction is to adjust what can be a rigid mindset. As a young scientist, you develop diligence in completing all of your tasks. This is because task completion brings you closer to your goal.

No doubt, you have memories from graduate school, college, and even earlier about the importance of completing your work in a timely fashion. In fact, many of us take great pride in the timely completion of a job. And of course, we remember our fellow students and acquaintances who didn't get their assignments turned in on time, waiting until the last minute to prepare for an exam, always struggling with completing a paper.

They were chronically behind and continually under pressure because of it. Your observations of these colleagues may have driven the notion of timely task completion even deeper into you.

The following three coping strategies when used together will allow you to defeat destructive reactions to this frustrating problem.

Coping Strategy No. 1

The timely completion of work can swell to a goal in and of itself, and you may criticize yourself when you are unable to meet your calendar's expectations. Falling behind, you sense that new required tasks whose arrival you anticipate cannot be completed on time, and the time to exploit new opportunities will evaporate. As the backlog builds, you fall irretrievably behind schedule and begin to fear that your career progress is beginning to stagnate.

If you link your performance to your self-worth, then intense days that spiral out of your control can rob you of an important sense of self-satisfaction and reduce your sense of value. This is a dangerous trap that can all too easily ensnare you, undermining your capability to react quickly to changing circumstances.

In fact, you may unintentionally build this trap yourself. On those days when your work has been particularly productive and fulfilling, it is all too easy to let your sense of self-worth soar. You're buoyed by your own performance. However, the price that you pay for this self-image inflation on good days is self-image deflation on bad ones. Thus, the performance–self-worth link on the one hand and increasingly tempestuous days on the other conspire against you, robbing you of the excitement and enthusiasm of your young career.

There are two important realizations that can reverse your reaction to these particularly irksome days when you're unable to meet your performance objects.

The first and most important of the two is to separate your value from your performance. Performance is an external metric that changes from circumstance to circumstance, day to day, while your self-value is constant. Linking your own sense of self-worth to your ability to complete your task list is like strapping your sense of self-value in for a jolting roller coaster ride. There will be dramatic highs and disorienting, stomach-churning lows. However, while your competitive instincts may respond well to the challenge of an ever-changing task list, you must buffer your sense of value and adequacy.

Consider the most expensive and valued diamond. Its high value is constant in the face of sun or rain, war or peace, prosperity or poverty. Regardless of its external environment, the diamond retains its worth. This is the unwavering measure that you should place on your self-value. Your sense of adequacy should be well supplied, remaining elevated, even, and steady.

By decoupling your self-worth from your performance, you're free to watch your day come apart without fear of self-condemnation. In fact, being free of self-condemnation, you can apply the best part of your nature to repairing, salvaging, or completely rebuilding the day from the ground up. You're unencumbered in its reconstruction because its destruction has not damaged your value.

Thus, a "bad day" not only challenges your organizational skills; it's also a barometer of your source of self-worth. If as your day unravels and you feel the early pangs of inadequacy, recognize that these first pains are a monitory that your source of self-value has been disturbed, an urgency requiring your immediate attention to correct. It is more important to correct this unbalancing of your own internal sense of worth than it is to repair your broken schedule or task list. The latter is transient while the former is permanent.

Finally, be aware that your sense of value is not just vulnerable on a bad day. Good days can be their own source of trouble. It is just as important on these rewarding days that you do not allow your self-worth to be affected by the temporary boost that comes from these favorable external events. The tighter you permit this link to become in the good times, the more difficult it is to break during the bad ones. Insulate your sense of self-value from the exhilarating feeling that comes from a day well spent. Consider the following:

> After his general theory of relativity was proven in 1918, Einstein found not just recognition in the scientific community, but intense public interest as well. The growing recognition of his intelligence and insight, along with the demonstration of the veracity of his findings delighted a war-weary world. Einstein's sense of humility, when combined with his harmless banter and genuine, childlike charm catapulted this heretofore obscure and quiet scientist to the epicenter of media attention. The statement "It's all relative" became a popular expression of the time. However, Einstein himself was not prepared for this onslaught of adulation. He didn't know how to react to the crowds that greeted him and was confused by the avalanche of interviewers that pursued him.
>
> During a dinner party one evening, he met Charlie Chaplin, a popular movie star who was also a cynosure of the time. Since Chaplin had been a famous actor for several years, and Einstein needed an experienced perspective on the media frenzy, Einstein asked Chaplin, "What does it all mean?"
>
> "Absolutely nothing," Chaplain replied at once.

Guard and protect your sense of self-worth from outside performance. Allowing it to feed and grow on the sweet but nonnourishing product of good days allows it to starve during the bad ones.

Guard and protect your sense of self-worth from outside performance. Allowing it to feed and grow on the sweet but nonnourishing product of good days allows it to starve during the bad ones.

Coping Strategy No. 2

After guarding your self-worth from the influence of your performance, take the opportunity to examine your chaotic environment that so quickly spirals out of control.

Despite my best attempt to lay out my activities on either a paper or digital schedule, my day almost never proceeds as I planned. The analysis that I thought would take one hour actually requires two, and I don't get to complete it. The meeting that was scheduled to last for an hour expanded to two and in addition spawned a second urgent meeting that took another half hour, itself generating thirty minutes of paperwork and telephone calls. This is followed by an urgent call from the budget office that embroils me in an intense discussion because of a misunderstanding by an investigator about a project's budget. In the meantime, a nervous colleague, anxious about the anticipated departmental debate about her upcoming promotion, would like to urgently speak with me about strategies and tactics that she might follow that would be the most influential.

Of course, all these events occur in the climate of intense connectivity created by incoming telephone calls and emails. At the conclusion of this day, I review the wreck of my schedule and find that of the five items I wanted to accomplish, I worked on only two. Additionally, I now find that new activities need to be frontloaded on tomorrow's schedule. But of course, tomorrow will be much like today.

Dizzying, chaotic work days are disconcerting. However, the problem is not the disorienting days, but my reaction to them. In fact, the days are exactly as they should be. It is my flummoxed reaction to them that requires adjustment.

The hectic times with their simultaneous and multidirectional activities are exactly the days that productive scientists are supposed to have. The forward movement generated by a fast-moving career generates unpredictable activity and chaos. Chaotic days are to be anticipated and faced not with stupefaction, disappointment, and self-criticism but instead with resilience, strength, and curiosity.

Unpredictable activities that force their way into your schedule are the hallmarks of a career in motion. Think of your livelihood as a flowing river, producing rapid currents and strong eddies as it splashes forward. Its movements are unpredictable, sometimes fearful, but there is the power behind it. Of course, one way to control this river force is to dam it up, converting it into a quiet lake. However, the placid lake, while peaceful, exerts no real force. Platonic and motionless, it shapes nothing but is itself shaped. That is not what you want for your career.

Forward progress requires movement that is translated into frenetic and unpredictable activities that don't make it onto your calendar, but nevertheless make their way into your activities. While chaos should not be produced for its own sake, it is the expected by-product of the natural good momentum of your own career development.

> **Begin your day by recognizing that the most important event that will happen to you is the one that is not on your agenda but one for which you must watch.**

Therefore, recognizing that predictable chaos is an anticipated by-product of your advancement, and that your sense of value is not linked to your daily performance, free yourself from the tendency to stop the chaos and instead just divert and shape it for the good of your colleagues and yourself. You need not feel guilty for not completing your task list for the day.

Begin your day by recognizing that the most important event that will happen to you is the one that is not on your agenda but one for which you must watch. A new opportunity that allows you to influence and shape the ideas and work of others or that provides a new path of progress will intrude on you and require your attention. You don't want to miss this merely because it was neither preannounced nor on your calendar.

Your calendar is not your career; the events that you and others create that swirl around you are.

> **Your calendar is not your career; the events that you and others create that swirl around you are.**

Coping Strategy No. 3

You accept the fact that an important part of your day will be caught up in important and unplanned activities. However, this needn't be your entire day. Insist that at least a small component of your day be in your control by taking a "productivity hour" for yourself. During this time, work on a scholarly, professional activity that you particularly enjoy. Spend some time considering the project that you'll devote this effort to. Luxuriate in this decision process. More than a thesis or dissertation, more than a collaborative effort, this work will be entirely yours.

There are several advantages that accrue to you if you're willing to find and fight for this productivity time. Since we have seen that time is such a precious resource, choosing to keep some of it for yourself represents an affirmative recognition that you can't just sacrifice your time and energies for a project or for other people all the time. You need attention and sustenance as well, and the best source of this is from you.

Second, time spent alone in quiet productivity helps to quiet the restless voice inside you that insists you should be able to apply your best efforts in a direction that you choose and not merely to meet an obligation. You have spent many years training for and have made many sacrifices for your ability to conduct science in a field of your choosing.

Asking for one hour out of twenty-four to work on a topic of your choosing is not an undue demand. You may choose to expend this time on a book chapter that you have been anxious to start or a manuscript in which you'll explore a topic that has been inadequately developed in the literature. It may be spent on the design of a new device. You may work on a computation that has been bedeviling you. You yearn to work on these issues, but you just can't find the time. The issue is not *finding* the time—the issue is *taking* the time.

In addition, a source of difficulty in collaborative research efforts is the decision-making process, a process that at its best is based on compromise. The investigator team that's working with you on the project will not be able to incorporate all of your good ideas. Some of them will be rejected. Your independent "productivity time" serves you well here. Your ideas that are set aside by your colleagues are available for you to freely pursue during this time.*

Finally, your secured time may be the source of independent productivity. Relying on collaborative effort for the evidence of your experience and expertise can be a problem if the effort fails. After you have invested many months in the project, the program may fail for a variety of reasons. The funding source may dry up. The principal investigator may leave. Advances by competing researchers may negate the purpose of the project. A senior author may have a very leisurely approach in writing the project's results. These are beyond your control.

---

* I have to confess that my ability to explore some of these rejected ideas of mine actually makes me easier to work with in a group. I may lose a discussion with my collaborators on the value of a particular contribution that I had hoped to make in a joint effort. This defeat is easier for me to accept if I know that I can use and develop the idea in work that I am carrying out on my own.

However, by having a second source of productivity, you're not completely attached to the productivity of the group project. Your secured time may be an unanticipated wellspring of productivity by which your career progress can be measured.

Therefore, find a time and create a reliable atmosphere in which you can quietly work. The duration of this environment may be only one to one and a half hours a day. Intelligently choose the time, but insist on finding it.* During this time, answer no phone calls. Turn your email client off. If people find you in your office and attempt to root you out, then go to the library. If you have an assistant, be sure to know that you're only to be interrupted in an emergency. During this time, totally immerse yourself in the project of your own choosing.

Three Coping Strategies for Chaotic Days

1. **Separate your self-worth from your performance. Allow your day to come apart without fear of self-condemnation.**
2. **Understand that unpredictable days are the hallmarks of a career in motion.**
3. **Find a time and place where you can quietly pursue your own scientific agenda without interruption for a short time every day.**

Free Time

Occasionally, I will have a blessing of time. A meeting may be canceled. A hearing may be postponed. Because of a natural disaster, an abstract deadline may be delayed. Suddenly, I find I have time that I thought would be full and off limits. I have "free time."

In a period where time is our most precious commodity, "free time" is something to be carefully considered and not squandered. I would encourage you not to spend it by pushing ahead on your schedule. Try not to use the time to just accelerate your calendar.

If you have planned your schedule carefully, you shouldn't need to put your schedule or yourself into overdrive. Instead, do something different. Spend some of that time with your family. Visit a relative. Make

---

* My secured time is from 6:45 AM to 8:15 AM on workdays and from 5:30 AM–7:30 AM on the weekends.

plans to get together with a friend. Visit with some colleagues to learn how they are doing.

Unanticipated free time doesn't automatically translate into additional time to work. Working hard doesn't mean working all the time.

In an environment where there will always be more work to do, forget about getting a jump on the labor. You will never catch up. Your usual diligent attention to your schedule will handle the productivity. Use the free time to spend on other important nonscientific activities.

Accountability

In the complicated and unpredictable mix that represents your day, you'll experience setbacks, occasionally missing a deadline. In these circumstances, take full responsibility for the setback that you caused your project.

When I have missed a deadline, I have found that the simplest, honest comment that serves me the best is the following:

> I missed the deadline. I'm sorry. It's all my fault, and I apologize.

Time and time again, this comment has produced more goodwill and more support for my efforts as an investigator. I may follow that comment with some of the specifics about why I missed the deadline. I also work out a new timeline that I should be able to make.

However, a clear, unambiguous statement of apology and responsibility conveys the message to the investigators that this is an issue for which I am accountable and that I will correct.

We will see that an additional advantage of an apology if you are a leader of a team is that it protects your team members from outside acrimony. If your team consists of junior members, they will feel raw and vulnerable when a mistake has been made. Taking full responsibility for it allows them to proceed with their sense of value intact and also learn from your example.

I was the principal investigator for a coordinating center that helped to design, organize, execute, and publish clinical research results. Approximately, midway through the project, we hired an assistant who was tasked with helping to update our adverse-event reporting systems. During her review, she found three of our published papers that appeared in prominent journals of which our investigators were very proud that contained mistakes in safety reporting. Specifically, the number of events reported was incorrect.

Cell therapy was and is a controversial field, and we as well as the community were always concerned about any adverse effects associated with this therapy. Precise and accurate reporting were key; mistakes could mislead.

But there was also the question of confidence. How could the coordinating center be trusted in the future if these events had been permitted through?

We also had adversaries of the coordinating center. With this new information about our mistakes, they could make arguments diminishing our authority.

The complexity of this circumstance tempted me to deflect responsibility. It would have been easy to do, the pressure of work, the mistakes of a subordinate were each likely candidates. However, I chose the following tact.

Our team did a full review to discover the scope of the problem. We corrected our procedures to remove the possibility of this particular mistake.

In addition, we prepared errata to be submitted to the appropriate journals.

I explained the problem to my director and department chair, saying that it was all my fault and that we were taking steps to compose and submit errata to the relevant journals.

I spoke to the executive committee of the NHLBI-funded studies, explaining the problem. I again took full responsibility.

I then spoke to each of the principal investigators of the study privately and then to the steering committee as a whole, explaining the situation, taking responsibility, and laying out the corrective steps.

You never know how other people react privately. For me, it was a cleansing experience, and I received valuable advice and wisdom from senior people in the field on how to manage community reaction. Everyone, now fully versed on the problem and its solution, moved on.

Also, consider the reaction of my team. Imagine how they felt when they arranged a meeting with me to first bring this problem to my attention. They did not know how I was going to respond.

They couldn't know whether I would fly into a rage, or engage in histrionics, or even demote/fire them. They had the strength of heart to bring this issue to my attention even though they knew that they might be damaged by my reaction. Their courage gains them a full measure of the credit for the successful handling of this situation.

Miscalculations, errors in judgment, bad observations, and the play of chance are as inevitable as tomorrow's sunrise. The occurrence of ineluctable, unfortunate events doesn't determine whether you're a good versus a mediocre scientist. Your reaction to them does.

Becoming vs. Being

Having mastered the background knowledge base, be quick to take advantage of your mastery. Use it to serve the project. Make a full contribution to the project's effort. Be cognizant of what is going on in all aspects of the study and act of your knowledge. Cease trying to become an investigator and start being one.

Junior researchers commonly wonder when they can stop being junior researchers. The answer is that you're a full-fledged investigator when you perform like one. A mature investigator seeks all the ways that her knowledge base can contribute to the project then sees to it that her contribution is most clearly and constructively made.

When you're functioning at this perceptive and incisive level, you can drop the "junior" descriptor from "junior investigator."

Many junior investigators try to remain as quiet and as inconspicuous in a new project. While they may believe that there is safety in silence, in fact, there is only the false peace of being content with their level of knowledge (and ignorance), and the sham serenity of assuming that the problems of others in the project are not their problems.

This is a mistake. Specifically, by not being a full participant in the discussions of the project, by intellectually walling themselves off from and separating themselves from the ongoing activity, they are not testing their own knowledge base about the material. Remaining inconspicuous and quiet stunts the junior investigator's growth and prolongs his maturation period in the research grant.

Gain experience. It is experience that gives you the sensitive feel of action that in turn leads to solid and reliable intuition. Extend your influence in the project by identifying ways that you can work with coinvestigators for the good of the group effort.

Look for every opportunity to expose your education and background to others in a way that supports them. Not only will this benefit the overall research effort, but it will reveal critical weaknesses in your own developing thought processes, allowing you the opportunity to repair them.

It is, of course, important to avoid the rash statement and the thoughtless action. By the same token, avoid the mistake of the opposite extreme. While rashness should be avoided, irresolution should likewise be shunned. If you have something useful to contribute, then make the contribution.

My first role as an investigator was with a group of forty cardiologists who had designed and were involved in the early stages of the execution of a large clinical trial in heart failure.

My transition was a quick one; I received my PhD on a Friday and was involved in discussion with this group the following Monday. A few days later, we were planning a presentation of the initial experience of the study. I knew relatively little about this modern therapy for heart failure that was being tested and did not know the investigators at all. However, this group recognized that I was working to master this material and patiently answered each of my "dumb questions."

Six weeks later, it was decided that I would present our data to an audience of physicians. There is no doubt that I was overwhelmed by the prospect of presenting to a roomful of cardiologists who knew more about the disease and its modern treatment than I did. Nevertheless, I was also conversant with the materials and felt that I would be supported by my experienced colleagues. In any event, one thing was sure. Regardless of the outcome of the presentation, my knowledge base on the subject matter would be much stronger after my appearance because of all the effort that went into my preparations. It only remained for me to control my own nervous faculties and energies in order to give the talk.[*]

> **While rashness should be avoided, irresolution should also be shunned.**

Engage in the Action

How you conduct yourself in a collaborative project is how you'll be treated by your colleagues. Professionals engender professionalism in others. However, a professional demeanor is not a rigid one. It is not isolationist, but participatory. So while you must always be appropriate, appropriately plunge in and take part in the activities of the group.

---

[*] Dealing with presentation anxiety or "stage fright" is the topic of chapter 21.

Your activity and participation in the scientific exchanges with your coinvestigators produce several good products. First, it is good for the group. If your coinvestigators are to benefit from your knowledge and training, then they must hear from you on a regular basis. If your area of expertise is complicated and/or technical, you may have to elaborate on the details repeatedly until they understand and can begin to integrate this knowledge with that of their own.

In addition, engaging in the brisk give-and-take of the group discourse allows you to educate others about the utility and tools of your own field. Experience in these dynamic dialogues generates an intuition that will be invaluable in your writings and presentations.

Be Persuasive

There is much to be said for the art of debate in science. An active discussion between scientists who have different points of view can be an enlightening experience for each of them. Commonly, these are short exchanges that occur and disappear like eddies in a stream. Be ready to inject important, pivotal perspective when they occur and to defend your statements.

You may find that on occasion, you'll sharply disagree with other coinvestigators (some of whom are more senior than you). Don't begin the discussion until you're clear in your own heart and mind that regardless of the outcome, the ensuing discussion will not change your value.

Since your self-worth is separate and apart from your outside experiences, you can fully engage all of your faculties in the discussion without fear of failure since failure will produce no long-lasting damage. Living in the affirmation that your value is separate and apart from your performance keeps you centered.

The central characteristic of a good persuader is not so much her ability to speak as it is her ability to listen and to understand. Many times, the discussion can have at its root a misunderstanding. Before the discussion proceeds, first rule out any misunderstanding by listening carefully to the other argument.

Since your self-worth is not at stake, you can absorb the opposing point of view with no harm to you. Examine it carefully, and ask questions about it until you completely understand it. Give its supporter every opportunity to correct and clarify it. Demonstrate to them by your questions, your review of their point of view, and your tone that you respect their position.

Living in the affirmation that your value is separate and apart from your performance keeps you centered.

Try to avoid any brisk point-counterpoint engagements until you fully understand the opposing position. In fact, it may take all the available time at a meeting to discuss the opposite point of view. Do it in an unrushed fashion. Go out of your way to both learn and demonstrate your willingness to listen. Also, do not be quick to anger.

Respond kindly to a harsh word. Your role on the team is solid. Admit your mistakes or misunderstandings readily and easily.

Many of the "debates" in science are based on different sets of assumptions. At the end of this part of the discussion, you'll know whether this is the case. If not, then begin your response, educating him about your perspective on the issue. State your point of view in a language that he can understand. Don't lapse into technobabble. Speak slowly, carefully, and clearly.

If it would be useful for your fellow discussant's understanding, provide some helpful, clear examples that he will be able to accept. Focus on the concerns that he raised, and overcome them with clear, evocative language. Use a point that he made to bolster your own position. Rely on your knowledge, training, and expertise to clarify your position.

Simple reliance on these tools can provide the basis of a profoundly convincing argument as in the following hilarious example:

> After stopping for drinks at an illegal bar, a Zimbabwean bus driver found that the twenty mental patients he was supposed to be transporting from Harare to Beltway Psychiatric hospital had escaped from his bus. Not wanting to admit his incompetence, the driver went to a nearby bus stop and offered everyone waiting there a free ride. He then delivered the passengers to the mental hospital, telling the staff that the patients were very excitable and prone to bizarre fantasies. The deception wasn't discovered for three days.[*]

---

[*] The Darwin Awards, 2003.

At the conclusion, with both parties understanding each other's points of view, the correct position, or a compromise position, may be self-evident to all.

Sense of Humor

Persuasion is not a science; it is an art. Being persuasive doesn't mean that you always argue as though you're immune to criticism. The art of being persuasive requires that you demonstrate your own vulnerability. One of the most satisfying and useful ways to do this is to retain a sense of humor about your own work. Be easily entertained by both your own comical missteps and the comical errors of others. After all, there is no scientific tenet that denies us the right to have fun being colleagues.

One common mistake that many junior investigators make is that they embrace their own work with almost religious solemnity. Remember that your self-worth is not based on your work thesis or product, but instead finds its roots in an internal source that is protected and buffered from the daily swirl of your activities.

Look to strengthen this foundation, inspecting it for any weaknesses or damage that the experiences of daily life can inflict. A useful indication of a healthy self-assessment is the ability to make fun of yourself. The more serious the argument, the greater the need for a sense of humor.

This last comment may be surprising, and actually, some scientists may be insulted. However, having a sense of humor about your work does not mean that your work is frivolous. The statement's justification is that while your work is very important, it is not all-important.

And while you derive a deep and lasting satisfaction from your work, it should not be the source of your self-justification.

Don't be a fanatic. Consider the following:

> Most of the battles during the US Civil War in the border states were not engagements over the titanic issues of the rights of men vs. the rights of states, but were instead smaller fights between neighbors.
>
> In one such affair, a small but intense fight broke out between Union and Confederate militia from the same town.
>
> After this engagement in which several men were wounded and killed, a hush fell over the field of fire.

A few minutes later, the young Confederate soldiers heard new loud noises coming from their surviving adversaries. The youngest of the southern soldiers, unsure of the reason for this strange grinding mechanical noise, could restrain himself no longer.

In a shrill, clear voice that was heard across the small battlefield, he called out, "Hey, Yanks! What're you up to?"

The reply from the Union side of the line was brusque, official, and all-business. "Shut up, reb, it's a military secret!"

A moment later, the southern soldier responded "Awww . . . we know that. But surely, you can tell *us*?"

Loud laughter erupted on both sides of the battle line.

Although friends aren't always friends, enemies aren't always enemies either. If these men could retain a sense of humor in their deadly killing business, surely we can find something amusing in our own more gentle affairs.

As a statistician, I commonly must work with investigators who enthusiastically accept my presence on their team. They know that their work product will ultimately be reviewed by a statistician before publication, and in order to avoid any lethal criticisms at this final level of review, they engage my services early in the development of the research project.

My colleagues accept the fact that I will make an important and necessary contribution to their research program, yet the contribution is one whose details they do not understand, generating a level of suspicion.

At the end of a mathematically intense demonstration of an analysis that is required for their work, I sometimes encourage my nonmathematical coinvestigators to keep in mind that despite the rectitude of the mathematics,

> statistics is like a bathing suit. What it reveals is interesting, but what it covers up is critical.

What makes this so funny is that my statement reflects their thought processes. They would have never thought to frankly articulate these to me because they feared that I would take offense at their perceived denigration of my field.

The fact that I can make fun of my own area of expertise demonstrates that to some extent, I appreciate their perspective. By sharing this concern, I can be better trusted to guide them through this mathematical wilderness.

Gain a Sense of the Transient

In science, we seek to add to a rapidly growing body of knowledge. However, each of our contributions, while not negligible, can be relatively small; and almost all serve a temporary role.

I have had the opportunity to make some useful contributions to statistical methodology. I devoted important energy to this work and am proud of it. Yet I would be surprised if these innovations and advances were in active service fifty years from now. Furthermore, I have had the opportunity to work on the development of new and important medications in the treatment of cardiovascular disease. These medications are helping people today. Yet I would be amazed if these medications were still being prescribed fifty years from now.[*]

As a junior scientist, you must commit your effort to work that is timely, but who among you can say that time will not pass your contribution by? A sense of humor about your work helps you to keep this in balance.

Respect Your Own Judgment

Junior scientists frequently make the mistake of disrespecting their own good judgment. Sometimes they prematurely give an opinion about a scientific issue before they are prepared to simply because they have been asked to provide one and they are "on the spot."

Other times they do not provide an opinion because they do not know how to formulate one after a review of the data. Since criticism commonly follows the delivery of an opinion, the junior investigator who hopes to shun criticism chooses to remain quiet.

It is perhaps all too easy for you to debunk your own judgment. After all, after living with it all of your life and being fully cognizant of the difficulties it's caused for you at times, you're painfully aware of its shortcomings.

---

[*] After all, how many of the medications that were developed with great fanfare and acclaim in the 1950s are still used today to prevent heart attacks? For example, at that point, a healthy diet was a daily regimen of breads, eggs, bacon, butter, meat, and milk, a diet that many would criticize today.

However, others may have a very different perspective of your judgment. While you may not value the words that you utter, others can hold them at greater value. Do a good service not just for yourself, but for others as well by carefully formulating and then carefully articulating your opinions.

If you're not ready to make a judgment, simply say so. While you're formulating your judgment, make sure that it is a complete judgment, not simply a narrow, scientific one. Expand its base.

For example, while there is no doubt that an important part of the analysis of the question at hand is scientific, is there no room for compassion? For humility? For respect for work that is not your own? For the desire to strengthen the work of others?

Integrate these perspectives as well into your scientific analysis so that your evaluation reflects the important nonscientific as well as scientific perspectives. Cultivate your judgment. Think about it, examine its basis, and challenge it. When your opinion is fully formed, practice providing your opinion lucidly, unhurriedly, and gently. Also, be prepared to defend it without fear. Criticism does no harm and provides the greatest benefit to the scientist who has a healthy sense of self-value.

Keep in mind that people who don't know you very well tend to believe that you're putting good thought and effort into what you're saying. A comment that you toss out cavalierly or thoughtlessly will likely be seen by strangers as being representative of your point of view. And of course, once the statement has been made, it cannot be unmade. When you speak, keep in mind that more people may be listening to you than you know.

Manage Your Email Content

This latter point is important with institutional email. The contents of an email message that you write can help to elevate you or lead to your ridicule.

Once you send an email containing an opinion, you have lost control over the dissemination of that message. Offhand comments that you make in an email to a friend can be routed to the attention of others.

Emails are commonly sent to the wrong person. You may have perhaps intended it for one person, but sent it to many more. Additionally, the person to whom you did send it forwarded your message to others, perhaps by mistake, perhaps not. A message that you sent that you hoped would be harmless can, through a circuitous route, wind up on the computer or smartphone of your supervisor. Sometimes they wind up in the office of the dean or the chief executive officer.

Be wise about your email. Compose your email with the idea that it could be read by your principal investigator, the state legislature, or a regulatory commission and not cause you embarrassment. Be sure to read each email message you send two or three times to ensure that it says exactly what you mean.

Never think that you're too tired to do work and therefore you'll catch up on your email. Sending messages in this state is as bad as sending them under the influence of alcohol. If you're too tired to write representative defensible email, then you're too tired to read any. Shut your email client down and find something else to do.

Your judgment is held in high regard by others. Make sure that you accord it the same status as well.

---

1   Monahan BP, Ferguson CL, Killeavy ES, Lloyd BK, Troy J, Cantilena LR Jr.(1990). Torsades de pointes occurring in association with terfenadine use. *Journal of the American Medical Association*. 5;264(21):2788-90.

14

Administrating Investigator

In reality the easy part is writing the grant proposal. The truly hard part is conducting the study! Most people are not prepared for the difficulties of hiring/firing and managing staff, managing budgets, balancing grant management, teaching, administrative and other responsibilities . . . There is a whole other side that scientists are not prepared for.
—A frustrated junior scientist, 2003

The Proposition

Devote regular and frequent time to administrative duties. For the first month of employment as a junior scientist, spend 50 percent of your time working through administrative issues. In the next four months, reduce this effort to 25 percent of your time. Beyond that, reduce the time that you spend in administrative activities to between 10 percent and 15 percent of your effort. When you become the principal investigator of a research project, increase your administrative time commitment to between 30 percent and 40 percent.

It is my hope that at the conclusion of this chapter, you'll be convinced of this statement's worthiness.

Administrative Oxygen

The underlying theme of this text is that professional and character development, not productivity, are the stars by which the investigator should guide their career. Therefore, since the unremitting pursuit of productivity is not the only metric that determines the scientist's worth to either themselves or their institution, productivity can temporarily be set aside, replaced by other worthy considerations. Chief among these is administration knowledge and responsibility.

There are many administrative activities that must occur in order for us to carry out our research and pursue our careers; we frequently relegate these activities to the background. Our buildings are available for us to use. Our offices are provided, always available, and amply heated and lit. The telephones, monitors, and elevators work. Support staff get paid in a regular and predictable fashion.

Although we commonly consider such activities as autonomic, occurring in the background without effort on our part, their proper and consistent execution requires the dedicated efforts of many. In some sense, administration is like oxygen. Whether we acknowledge it or not, whether we like it or not, we need it and therefore take advantage of it.

Experience vs. Expertise

Before we begin, we have to remove from our consideration one unhelpful notion; that is, that scientists can work without administrative experience.

This is a canard. Every scientist gains administrative experience. The real issue at hand is whether the scientist can gain good, useful, and industrious administrative skills on the one hand or frustrating and counterproductive incompetencies on the other. For many, experience is simply repeating the same mistakes. Therefore, as a junior investigator, you're not interested in gaining mere experience, but instead, should focus on acquiring expertise. You're interested in learning the right lessons from your past and applying these lessons in the future.

Administrative diligence is commonly the least appreciated but the most useful and rapidly acquired skill that the investigator can acquire. Attention to administration can facilitate your intercollegial interactions, speed your travel arrangements, allow you to purchase necessary equipment, and help you to obtain support for your class. Administrative attentiveness is the means to these productive ends. In essence, the successful application of administrative efforts allows you to more quickly bring needed resources to bear in order to carry out your work.

Surroundings

As a junior scientist, you may feel (or sometimes be made to feel) that you have to jump into the middle of your scientific work right away. This means that you'll work with a small number of individuals, talking and communicating with them regularly.

If your work is technical, then you'll in all likelihood be required to focus your attention on small but important methodological details. However, it is very useful and instructive to simultaneously obtain a larger perspective on your institutional surroundings. Learning how your school, university, corporation, or institute views you and your work product can provide a guiding point of view, updating your perspective on your own work.

Institutional Perspectives

Many institutions periodically review the productivity of each of their scientists, evaluating their caliber and quantity. This evaluation may consist of an assessment by a department chair or division leader on the one hand or a committee review by peers on the other. Commonly, evaluations for important promotions are carried out externally at a state or national level.

Many junior scientists are surprised by the negative reviews of these superiors, whose lackluster comments fly in the face of the young scientist's own convictions. He believes that he's been industrious and that this diligence has generated a worthy product, for example, new technologies or published manuscripts. Nevertheless, the reviewers were unmoved by this scientific output. Even though the junior researcher came to believe in the importance of his work, the institutional perspective of his activities was quite different.

One possible explanation for the concerns expressed in the evaluation was the sense that the junior scientist's work was not really contributory.[*] The reviewers did not deny the productivity; they simply don't think that the productivity was on track. From the assessors' perspectives, the contributions were neither helpful nor in accord with their goals and expectations for the junior scientist.

The reviewed scientist may have anticipated this reaction if he had understood his organization's perspective on his own work. However, this

---

[*] In a small minority of cases, these reviews may have been meant to do damage and be personally disparaging. This is an issue that we will discuss in a later chapter.

understanding was missing because, like most of his peers, he was just too busy working to seek it out.

As an investigator, you may or may not agree with the institution's current priorities, but you certainly have to know them. Most commonly, this means taking the time to learn about and absorb these goals. If you work for a corporation, then a clear understanding of the agenda of the company and how you fit into that corporate schema can be very instructive.

A junior faculty member at a university is best served if she understands the mission statement of the school and how that mission statement is translated into goals and objectives for faculty activities.

Knowing the institution's ideas and plans for your own work can be an illuminating and early guide as you begin to shape your own research agenda. This knowledge can help you to avoid surprising reactions by distant superiors to the reviews of your work, superiors who are more closely aligned with the institution's ideas and plans than you are. Similarly, understanding institutional goals and objectives can remove misunderstandings of the importance of your own work, misunderstandings that can damage your career advancement.

Local Environment

Assume that you'll work forty hours a week (clearly an underestimate for most of us) for fifty weeks a year over forty years. A quick calculation shows that your career will consist of eighty thousand hours of work. Choosing to take only 15–20 of these eighty thousand hours to learn what it is that the people who support your department actually do can pay handsome dividends for everyone, dividends that far exceed your relatively small-time investment of time.

It is a truism that most institutions that hire scientists recognize the support these professionals need. This effort includes editing support, imaging management, scheduling support, and supply procurement. To this traditional list, we can add information technology support, including but not limited to computer hardware support, email support, electronic security, and software support. These echelon units do not exist merely for themselves. They are provided for the purpose of assisting you. Their raison d'être is to support your work. Yet these resources are squandered if you do not know how to use them in the right combinations that will amplify your own efforts.

The best way to understand how these units can shore up your work efforts is to appreciate (1) what it is that these support units provide, (2) how their product is offered, and (3) who pays for them. This requires an investment of your time and effort. To be sure, some of this material will be covered in whatever orientation that your institution offers you as a junior scientist. However, sometimes efficiently run orientations can be disorienting because they cover so much relatively unrelated material so quickly.

Therefore, consider the following strategy in order to more properly understand the role of these assistance groups. Several days after the official orientation has concluded, conduct your own unofficial reorientation. Revisit the departments to learn in some detail what is it that they do and specifically how that work can support your own productivity goals and objectives. Most importantly, be sure to add the names, phone numbers, and email address of these support people to your contact list. Take the time to add a note about what it is that these people do.[*] Understanding what the various departments accomplish and how they interact creates a solid logistical foundation for your work.

This reorientation not only provides a more complete assessment of the roles of these individuals, but your second attempt at familiarization also generates a sense of appreciation in the support staff. People react positively when they see that the professionals they support have taken the time to express a genuine interest in both them and their activities.

> **Several days after the official orientation has concluded, conduct your own unofficial reorientation. Revisit the departments to learn in some detail what is that they do and specifically how that work can support your own productivity goals and objectives.**

Learn how these individuals are financially supported. Remember that their departments, like all institutional divisions, have to justify their work. If you choose to use their services, you should acknowledge their support appropriately. If you're applying for a grant, then you may have to

---

[*] I have found that while I can remember an individual fairly easily, I unfortunately will quickly forget what it is that they do. A brief note embedded in their contact information refreshes my memory at once.

include a specific monetary amount in the grant. Alternatively, you may be asked to complete a job request that would serve as documentation of your use of this service. The best way to learn how to provide appropriate financial support for the unit is to speak with the people whose services you require.

Occasionally, you may be moved to write a letter to the director of your institution or to the dean of your school, formally acknowledging the pivotal work of support staff for your product. Certainly, this effort would take more of your valuable time. However, in uncertain economic periods, when support staff are commonly among the first who lose their positions because they are "nonessential personnel," an honestly expressed show of support from a promising scientist can have a beneficial impact far beyond the twenty minutes that it takes to compose the letter.

Be Familiar with the Rules

Perhaps there will soon come a day when scientists and their support staff can function in an unfettered and regulation-less environment, but today is not that day.

Whether you're working at a private institution, for the government, or in academia, you can be sure of two things: (1) there are rules where you work that attempt to set the parameters of interaction and behavior and (2) the number of those rules is growing.

The reasons for the growth of these rules are multitudinous and beyond the scope of this discussion. The simple message here is that you'll need to pay some attention to these rules.

You may not be aware of it, but in all likelihood, part of the agreement or contract that you signed when you agreed to work at your institution obligates you to follow all of its rules.

Although you certainly have an idea about the caliber of professional conduct, you may be unaware of critical details. What are the rules for computer use? Can you use official institutional stationery for consultative work? Can the institution scan your computer for inappropriate material (and by the way, just what is the definition of inappropriate material?)? What constitutes correct and incorrect use of email? How should actual and perceived sexual harassment be handled? How should disputes about research ethics be mediated?

Knowing the answers to these questions can have a direct and personal impact on you. In a controversial situation that arises involving your activity and conduct, you may have behaved and acted correctly.

However, even though your behavior was well motivated, an important metric that the institution will use to measure your performance is whether you followed the institution's rules. This may be the final arbiter of your performance in a controversial area.

You don't have to memorize the rules; just have them handy. US Supreme Court justices who make complicated and critical legal determinations are, of course, expected to know the US Constitution. However, they don't memorize it, instead commonly carrying small inexpensive copies of this guiding document with them for easy reference.

Since the justices of the highest court in the land don't memorize the highest law of the land, you should not feel compelled to try to memorize your institutional rules. However, like the justices, you should keep a copy of these regulations within easy reach. At some institutions, these rules are in electronic format on an institutional website. If so, consider bookmarking them. If their size is not overwhelming, consider placing a copy of them on your smart device. Remember, you don't have to commit them to memory. You simply have to be familiar with them and know where to find them when you need them.[*]

"Not for Me!"

Before proceeding any further, we need to clearly face one issue. If you're named as the principal investigator on a research activity, be it a grant or contract, then it is yours (and no one else's) responsibility to administer that project. In order to be a productive scientist, you must become a competent administrator.

This can be a difficult responsibility for junior investigators to accept. You probably have no formal education or training in business administration, and most junior scientists have never run a business. We have spent most of our productive time becoming scientists, not attending business schools.

Therefore, lacking the knowledge, training, experience, and expertise to manage a business enterprise, they balk at the prospect of accepting administrative responsibility for a project. Junior investigators are commonly unable to learn the important coping skills that they must

---

[*] Keep in mind that in the dynamic environment that the workplace is, rules can change over time. Therefore, you need to examine them periodically (say, yearly) to make sure that important alterations have not occurred without your knowledge.

master in order to successfully run a grant because they cannot purge themselves of the notion that running the grant is not their affair.

Because these researchers think that administration is not their business, they refuse to open themselves to learning the skills and abilities that they must master in order to successfully supervise the grant. In the young researcher's mind, she sees herself as the scientist, not the grant supervisor. The job of administration, so she thinks, is someone else's job, perhaps anybody else's job, but certainly not her job!

The problem worsens for these new principal investigators as they are first approached by and then pursued by important and unresolved administrative matters in which they believe that have no stated interest and no real concern. These scientists find that their days are increasingly consumed by increasingly frustrating supervisory difficulties. Because they are never really handled successfully, these administrative problems rise up like weeds to strangle the scientific content of the grant. It is no wonder that these junior workers find the entire grant experience a messy and discomforting affair.

To some, the reaction of the junior scientist to the specter of administrative management is like the response of a new and naïve parent to the newly discovered difficulties of raising a child. Much of the work is unpredictable, messy, and sometimes painful.

However, it is the rare parent who does not divest themselves of their naiveté about parenting for the sake of their child. Good parents learn everything that they can learn and do anything that they can do for the sake of their children. Leaving their preconceived notions behind, they plunge (or are thrown) into the problems of child-rearing, using everything they know and can comprehend in order to raise their child right. Their responsibility is not to be perfect, but to simply and in an open-hearted fashion do the best that they can know and learn to do. As a junior investigator, you're better served by this model than by one that is based on developing a self-serving husk that by constraining your attitude stunts your administrative growth.

Effective Communication

The major purpose of communication is to influence the thinking and therefore the behavior and actions of others. The key to effective communication is to make the person with whom you're conversing at ease. When we are at ease, we are open; and by being open, we are most likely to hear and understand the message that is being transmitted to us.

Receiving the message in this spirit permits us to understand it, absorb it, and allow it to influence and guide our behavior and activities.

Sometimes the message is a short, but important one, for example, "you're not doing your job." This is an example of a difficult message for a worker to receive from you as the principal investigator. Therefore, in order to communicate this message effectively, you'll have to spend the time to put this person at ease. Only by being at ease can they accept this message and begin to make the adjustments in their behavior or conduct that would allow them to repair their shortcomings.

Thus, in communication, most of the actual conversation will be to ensure that the individual is comfortable. This doesn't mean the conversation must be full of platitudes about the weather or sports teams. Instead, these difficult conversations must use the words that convey the true sense of the topic; that sense is that you have the deficient worker's best interest at heart, are willing to spend some of your time working together with him to improve his difficulty, and that you believe, after coming to know the worker, that he can overcome his shortcoming.*

Since you do not mean to threaten the worker, the nonthreatening language and vocalisms needed to convey your information about his job performance to him should be within your reach.

Having invested the time to learn about your individual research team members, you're now in an excellent position to know the best words to say to them. There are, of course, various venues of communication. I would recommend that you use the most complete form of communication that is possible. For example, when the choice is between a face-to-face conversation and a telephone discussion, choose the face-to-face conversation. If the choice is between a telephone conversation and an email, choose the telephone conversation. These choices are the most time-consuming, but they are also the most expressive. This is because much more information is supplied by being

---

* Sometimes even this approach does not work. A young private in the Confederate Army during the US Civil War, when caught in an act of insubordination, was told to report to the local military tribunal for a hearing. When he reported as ordered, the disobedient soldier was shocked to learn that the tribunal for that day was not a low-ranking officer, but the commanding general of the army himself, General Robert E. Lee! When brought before the general, the teenaged private began to quake with fear at being disciplined by so distinguished a leader. General Lee, recognizing the fear in this young soldier, quietly said, "Don't worry, son. You'll get justice here." The private promptly replied, "General, that's exactly what I'm afraid of!"

able to read a person than by simply hearing what a person says. While attitudes can be communicated verbally, they are much more quickly telegraphed through face-to-face conversation.

Of course, much of your communication will be by telephone. Effective communication by telephone requires enhanced communication skills. Since the conversationalists will not be able to see you, they must rely on both the words you use and the sound of your voice to convey your message.

Keep any edge out of your voice. Try to speak slowly so that everyone can understand what your message is without feeling that you're rushing through it just to have it over with. Use kinder language than you might use in person. Again, because the person cannot see you and they must focus more of their attention on the words that you use, you must choose these words most carefully.

Try to avoid "dashing off an email." I don't always succeed, but whenever possible, I will revert to telephone conversations rather than email. There is no question about the superior efficiency of email; however, in my view, this efficiency is not always an advantage. Short, terse messages that are rapidly conveyed are not the point of communication.

The important communication between scientists should be more influential than efficient. When using email, try to make it evocative, that is, to actually contain the complete and exact sentiment that you wish to express. My rule of thumb is to overwrite an email, being explicit about my sentiments, hoping that by being more evocative, the sense of my reaction will be clearly communicated.

Recently, I had the opportunity to use email to correspond with a colleague about a paper that he was writing for our research team. Since he was away from the office and I did not have his traveling phone number, I composed the following first email draft:

Daniel

Read your paper. Here are my comments:

1. Introduction should be more focused on the scientific question.
2. Methods section doesn't mention the exploratory nature of the work.
3. Results incomplete—please include subgroup analyses.

This portrays the true caliber and description of my response to the manuscript draft, and the press of time may compel me to send this message.

However, its brevity is the problem. If Daniel receives the first draft receiving only my stark suggestions, he can only guess at my motivation. The message does not answer the following important questions about my motives:

> "Was my intent to be critical?"
> "Did I not like the manuscript?"
> "Do I not believe that Daniel did a good job?"

The answers to these questions are unclear to him because the message was "efficient" but unrevealing about my motivations. Daniel doesn't know that the press of time limited my comments because, after all, if I had more to say to him, he believes that I would have taken the time to say more. I therefore sent the following elaborative email:

Daniel,

How are you?

> You have worked hard on this manuscript, and its present draft reflects your consistent and patient efforts. You have explained some difficult concepts in a way that the reader is well educated about them without being overwhelmed.

When you get the opportunity, please consider the following:

1. The introduction has many fine components. It would be great if it could focus more attention on the scientific question we hoped to address in our study.
2. It might be best if we were clearer about that component of our work that was confirmatory (that is, actually answered a prospectively asked question) and that component that was exploratory. What do you think about adding a comment about this in the methods section?

3. I have read your results section, and it is a clear portrayal of much of our work. Can you consider adding a section that discusses subgroup analyses?

Give me a call if you have any questions. Thanks again for taking the lead on drafting a manuscript describing our controversial findings.

The only way to convey my complete reaction to the manuscript is to write a message that more fully and accurately expressed my total reaction. It needs to be evocative. Computers communicate tersely and efficiently—that's what they're designed to do. Humans do better with less efficient but more evocative interchange.

When communicating with your staff, be sure to have a clear and explicit component of your conversation, focusing on what they need for you to do that would allow them to move forward with their work. As the senior manager of this project, let your team know that an important part of your responsibility is to see to it that you're doing everything that you can do to ensure that their task go as smoothly as possible.

Computers communicate tersely and efficiently—that's what they're designed to do. Humans do better with less efficient but more evocative interchange.

Avoid keeping secrets. Everyone in the grant should have a complete understanding of both the long-term goal of the grant and the short-term grant objectives. They should also have the opportunity to comment on these objectives and the ability to suggest alterations in your plan. As long as your team is focused on their common goal, these suggestions will make your trial stronger.

When there are problems with the grant's administration, speak about them clearly and openly answering every team member's questions. Recall that the focus of attention in these activities is on not just the grant, but on you as principal investigator. How you conduct yourself is critical.

Conclusions

Make no mistake about it—you'll spend a good deal of your time carrying out administrative duties. While you have no choice about this, you can nevertheless choose how to spend this time. You can spend it consumed by an attitude of time selfishness that will stunt the development of your project. Alternatively, you can be governed by a spirit of time generosity.

Since administrative diligence requires this generosity, insist on and encourage the character growth required for you to accomplish it. This character growth begins with the recognition that your innate value as a scientist is not affected by your performance. We will have more to say about this in the next two chapters.

15

Mentorship

One of the first constructive administrative acts in which the junior scientist can involve herself in is the selection of a mentor. While the first reaction to this suggestion is commonly a combination of helplessness, indecision, and impatience, the dividends derived from this choice can be profound.

What is a mentor?

A well-chosen mentor can save your career.

A mentor is a senior scientist who becomes a friend, colleague, and confidant of the junior faculty member. The relationship is open and develops its own natural momentum. The mentor is willing to meet with the junior faculty member at a frequency dictated by the needs of the junior faculty and take time to understand the mentee's strengths, weaknesses, and motivations in an unrushed collection of meetings. This is to help ensure that when trouble comes, the mentor understands most of what they need to know about the junior faculty member to provide solid specific advice on the delicate matter.

The benefits of mentorship in professional communities are clear to all. These relationships are a perfect time for both the mentor and the junior member to sacrifice productivity to engage in a relationship that can have a beneficial fruit that endures throughout the career of both.

If your mentor is unwilling to sacrifice some productivity time for you, find someone else.

Your mentor can point out useful activities for you to become involved in and, just as importantly, help to direct you away from career hazards that you cannot see yet are easily visualized by your seniors. As a junior scientist, you're constructing and orienting your career compass to reliably point out the direction in which you should develop. This compass must be regularly tested, and it is the mentor who can provide this critical calibration on a regular and frequent basis. They can be a critical contributor to your character development.

A well-chosen mentor can save your career.

While the mentor's advice can be formal and is sometimes predictable, her counsel can be most effective and timely when it is based on solid knowledge about you. The more that the mentor knows and understands about your motivations and talents, the better she is able to serve you. The greater your investment in this process, the greater the benefit that you derive. Take the time to lay a solid foundation for their advice to you.

Meeting with Your Mentor
Mentoring meetings are the skeletal structure on which the relationship between the mentee and mentor is built. This support lattice must be stable and reliable because it is upon it that the relationship is built.

While there are many circumstances in which the hyperefficient use of time is important, meeting with mentors is not one of them.

The goal is not the meeting itself. Meetings are simply the surrogate product of the actual relationship. The meetings may need to be frequent and sometimes should be long. They are the opportunity for you and your mentor to share not just what you do, but more importantly why you do it.

Ask your mentor why she decided to focus on Lebesgue integration theory in health care research. Tell her why you are attracted to platelet aggregation theory. The relationship becomes easygoing and self-sustaining. Manuscripts can be shared. If the mentee is having a problem responding to the caustic review of a submitted manuscript, then a meeting with the mentor is in order. They can share stories of their victories and their defeats.

If the purpose of the mentorship is to share experience, then that comes from shared time, not a shared flash drive.

In order for the mentor to assist and counsel the junior faculty member, he must know of the mentee's defeats as well as their victories. The broader and deeper their perspective, the better the caliber of advice that they can provide.

While there are many circumstances in which the hyperefficient use of time is important, meeting with your mentor is not one of them. In order for the mentor to provide information that is tailor-made for you, they must know you, understand you, and gain a full and fair appreciation of your strengths and weaknesses. Therefore, avoid meeting with your mentor by email or by telephone conversation, instead choosing to visit personally.

Meet with your mentor on a regular basis, providing your own complete and honest appraisal of your progress and setbacks. In order for your mentor to assist and counsel you, he or she must know of your defeats as well as your victories. The broader and deeper their perspective on you, the better the caliber of advice that they can provide to you.

The junior faculty member requires a fair and critical appraisal. A mentor who always and only finds good things to say about them may be comforting but not constructive.

The junior faculty members should seek an honest critique from their mentor. All should work to create the cordial environment that permits the mentors to be comfortable in their candid discussions. The mentor should have the junior faculty members' professional well-being in mind and offer honest advice. In addition, the mentor should inform the junior faculty member on trends and developments in the chosen field of research and of opportunities for obtaining extramural funding and facilitate introductions to possible collaborators.

It is critical that the junior faculty member not feel that they are being pushed into a scientific direction that is beneficial to the mentor but does not support their own research trajectory. Keep in mind that you'll be most helped by fair, critical appraisal. Create the cordial environment that permits the mentor to be comfortable in making a fair and balanced appraisal of you. Of course, remember that while criticism of your professional development is essential, it should also be supportive and not destructive.

Crises in your career, like blizzards, droughts, or floods, are unavoidable. While you cannot always predict them, you can take some important steps to prepare for their inevitable arrival. The right mentor can be instrumental in a crisis.

During your time of trouble, you'll need the best support that you can find. You may be so overwhelmed by the difficult circumstance that for a time, you may be unable to rely on your own professional judgment. In this desperate situation, you'll need all the best and instructive advice that you can find.

A good, mature, and knowledgeable mentor can be a solid anchor to which you can cling during these tumultuous times. If you have stayed in regular and frequent contact with them throughout your career, he or she will have come to know and understand you. With the depth of this knowledge in full view, the mentor can provide timely and important advice for you in your time of difficulty.

16

Ethical Investigator

The Scope of This Chapter

You'll face an ethical dilemma as a scientist. Don't think that you are too smart, too clever, too connected, too blessed, or too good-looking to have to deal with an ethical threat. Like the next storm, an ethical challenge is coming, and it's coming for you.

Like character development, this chapter will help to prepare and strengthen you for this inevitable difficult time.

Here, we will not repeat the content of the many excellent general treatises on ethics that are easily available to you. Instead, our focus will be on the ethical issues that junior investigators must commonly address, concentrating on the unique vulnerabilities of these developing scientists.

Ethics

Ethical conduct is the demonstration of your respect for the work, contributions, and sacrifices of others. General discussions and treatises in ethics have been written,* but these rule books neither identify all

---

* Many sources are available on the web. A compendium of resources can be found at http://www.csu.edu.au/learning/eis/ethxonline.html. Concern for ethics, as might be expected, reaches down to the high school science level (http://www.nzase.org.nz/ethics/animal.pdf). There are ethics centers online (http://onlineethics.org/). In addition, there are now journals devoted to ethics (http://www.opragen.co.uk/).

good ethical behavior, nor do they delineate the complete universe of ethical misconduct. Rules may describe ethics, but ethical behavior is not a product of rule memorization.

Ethics itself is a way of life, derived from a lifestyle, flowing from an attitude, sustained by character development.

It may be more useful for you to think of ethics not as a collection of documents, but instead as a body of "living principles." These principles, like character development, growing and adapting over time, govern your interrelationships with others in science. These interactions are complex and interactive. They consist not just of actions but of words and not just of words but of vocalisms and "body language."

Codifying ethical behavior is like trying to photograph the wind. You observe its effects, but cannot observe the driving force behind these manifestations. Ethical conduct is not determined by protocol, but is governed and calibrated by an ethical character, itself undergoing continuous renewal.

Ethics is commonly reduced to identifying the right thing to do and then having the strength and courage to do it. However, it can be hard to know what the right thing to do actually is, particularly if one is not even prepared for the issue. Consider the following:[*]

> Germany was in ruins at the conclusion of World War II. A young American physician who arrived there at the war's end observed that Germany's urban foundation had collapsed. With no effective administration, communication, or transportation infrastructure, the ability of the German municipal governments to deliver food and medicines to its citizens had vanished. In this vacuum, diseases of poverty and poor sanitation arose to attack the old, the young, and the infirm.
>
> Particularly virulent was an outbreak of pneumonia, especially lethal among German infants. Fortunately, this infection was successfully treated with penicillin.

---

[*] This scenario was provided during a lecture while I was a medical student at Cornell University Medical College in the spring of 1996. I have long since forgotten the identity of the lecturer and so am unable to give him an appropriate attribution. However, perhaps it is credit enough that thirty years later, I still remember.

However, this antibiotic, then a relatively new drug, was in short supply.

Simultaneously, an outbreak of gonorrhea and syphilis broke out among young American soldiers who were consorting with prostitutes. Untreated gonorrhea produces a life of pain and misery. Untreated, syphilis would lead to insanity and death. The definitive treatment for each of these diseases was penicillin.

This young physician did not have enough penicillin to treat both cohorts, and he had to decide to which he would administer the drug. The German infants who had no responsibility for the 50 million deaths that the war produced would be its newest, last, and perhaps most tragic victims if the doctor treated the soldiers.

However, the GIs had not wanted to come to Europe and fight, but were compelled by the actions of a country that had given itself over to a tyrannical despot. Thousands of Americans had died or been maimed by choosing to fight in order to set other people free. They were not clerics; they were simply young men acting like young men commonly did. Were they to be punished by withholding treatment?

The physician did not have enough medicine for both. Who should he treat?

Many ethical people believe the German children should be treated without questions. Other ethical audiences insist that the GIs receive the antibiotic. Other ethical people, perplexed by the dilemma, recommend splitting the available antibiotic into two proportions, one reserved for the infants, the second for the soldiers. Rule books on ethics are of little value in these types of situations.*

Ethical Statement

While rule books can be useful, it is most helpful to have your own clear ethical statement. There are two important reasons to be able to articulate your own ethical point of view. First, you must have a point

---

* The physician chose to use the penicillin to treat the GIs. He said, "Every morning, outside my office, the young German mothers would line up, clutching their coughing and dying babies to their chest, and spit on me as I walked by."

of view in order to enunciate it. Your initial failings at the attempt may point out that your ethical perspective may not have evolved sufficiently, an improvement that requires your attention and careful thought. This requires reflection that comes from character development.

Secondly, you'll be required to clearly and cogently state and defend your ethical point of view when either your scientific work faces an ethical challenge or you're required to comment about the ethical issues surrounding the work of others.

Having already become accustomed to verbalizing your ethical point of view in a way that communicates your perspective in an informative, nonthreatening manner will be valuable. Work on developing a concise statement that summarizes your ethical philosophy.

For example, after much practice, I have come to articulate my own belief in the following two sentences:

> The ethical treatment of myself and others is based on the principle that we each have innate built-in value, regardless of our accomplishments or our treatment by others.
>
> This natural value requires that I treat the individual and their resources with respect and dignity, regardless of my opinion about their point of view, their accomplishments, or their standing in their field.

This philosophy has taken years to develop. I spend time in my character sabbatical thinking and working on it, testing whether it is worthy of my adherence, and if so, whether I am abiding by it.

With your personal statement in place, you now have a metric against which the ethic of your own daily behavior and activities can be measured. However, this metric must be constantly challenged. The challenges lie all around you. Since your colleagues will have an equally personal ethical philosophy that differs from yours, you may find that the conclusion that you come to in a particular ethical circumstance is different than theirs.

Innate Value

The importance of self-esteem, of prizing your own value despite the nature of external circumstances, is a core character principle for the scientist. In a manner of speaking, the recognition that you have great value regardless of how your colleagues or superiors treat you is the ethical treatment of yourself.

The ethical treatment of others is based on the principle that they have innate and special value, irrespective of their external circumstances or reputations. Good ethical behavior flows easily from your decision to accept, with approval, the belief that your colleagues, adversaries, critics, experimental subjects (human or animal) each have special and unique significance. Their worth remains unchanged regardless of how they are treated by others or how they treat you. This is the basis for reacting to them and their work product with deference and dignity.

Recognizing the innate value of others reinforces your drive and desire to respect them, their efforts, their time, their work, and their opinions. This is a very high standard and one that requires consistent effort and character development from you to achieve. However, this core principle of the innate value of yourself and of others provides clear direction for you as you gauge your relationships and interactions with them.

Undertows

It is unlikely that the reader will dispute the importance of moral rectitude and good ethical conduct. It is also reasonable to assume that your peers and colleagues shun unethical behavior as well.

At no time did a fellow medical or graduate student ever affirm that they wanted to be an unethical physician or that they aspired to become a dishonest scientist. Nevertheless, some of them have given themselves over to poor ethical conduct.

Examined from another perspective, senior scientists who are currently engaged in unethical conduct today started their career, in all likelihood, with the attitude that you have or had as a junior investigator. They did, as you do, cling to a firm belief in proper professional conduct. Yet they too have gone astray. A relevant question for you is, "What is it about me that will make me different from professionals who have come before me and who have lost their ethical way?"

Why do some junior scientists with the best and most wholesome of intentions ultimately engage in disturbing conduct? Part of the reason is that young workers are not looking in the direction of the ethical attack. As a junior scientist, you can't help but be attuned to sensational issues in ethics that may be in the news (e.g., falsifying data in a manuscript, or stealing authorship of an abstract, or withholding therapy from ill patients to make a research observation).

However, large ethical difficulties in oneself do not spring up de novo; they inexorably develop from character flaws that the scientist allows to go uncorrected. These defects, either hidden from view (or hiding in plain view), emerge to exert their overt influence during a time of duress, anguish, frustration, or fear. Thus, the scientist must consciously develop a talent of self-discernment through continued character development in order to recognize their own small vulnerabilities. Recognition of these flaws is the first step toward their removal or, more accurately, replacing these weaknesses with newfound strength.

> **Why do some junior scientists with the best and most wholesome of intentions ultimately engage in unethical conduct? What is it about you that distinguishes you from scientists who have lost their ethical way?**

The ethical nature of your character requires constant attention and must be guarded. However, recognize as a junior scientist that rarely will your personal ethics be overcome in one hurried assault.

Your boss won't storm into your office one day, interrupting your quiet deliberations, to demand, "Carry out the following flagrantly unethical task at once!" Instead, your ethics will be overcome slowly over time. This corruption, like the slow stream of poison sliding into a clear mountain lake, is a process that occurs over years and produces a predictably destructive and barren result.

The temptations and opportunities for unethical behavior surround us. Including a favorable but inaccurate summary of a preliminary experiment in a competitive grant application to strengthen the argument appears at the time to be the smallest of violations. The incorrect attribution of the ideas of a student is easy to get away with.

The promises of money, prestige, promotion, and grant awards come closer to reality if the investigator is willing to engage in a little unethical behavior.

We, as investigators, are surrounded by these temptations every day; like a strong undertow, these forces pull us away from our ethical base. If we are passive, this unethical undertow will sweep us downstream. Your vigilance through character development is required to detect these circumstances, and your affirmative energy is necessary to resist the temptations that they offer.

Damage to your ethics through enticements or distractions can occur in any career field. In some environments, the distraction is money.

Of course, everyone is concerned about money; and commonly, money is a good reward for good work.

However, in some environments, workers fixate on money. Although the ongoing central activity at the institution is supposed to be science, much of the conversation that take place is about money. Discussions of stock prices, estimations of the sizes of year-end bonuses, computations of the impact of a new product on one's wages, conversations comparing the salaries of different people can be ubiquitous and appear to swirl around you.

Even the most charitable and least material people can be carried off by this monetary maelstrom. Dissenting voices are drowned out.

Everyone in the organization agrees that ethics are important; however, if an outsider were to judge the organization's preeminent principle by the prevalent topics of conversation, the most important concept would be seen to be not ethics, but financial wealth. The atmosphere isn't seasoned with money; it's poisoned by it.

In other fields, the focus is not on money directly, but on publications. In a field where publications are intended to serve as simply a method of communicated scientific information, they can unfortunately become an end unto themselves. Colleagues discuss and compare the number of publications that they have. Researchers consider the advisability of working on projects based on the work's "publication potential."

Of course, publishing is important to the scientist, but the love of publishing (i.e., the idea that publishing has great value regardless of its content) blights the ethics environment.

Working and hoping to become successful in these environments can first disorient and then overcome your sense of ethics. As an investigator, you may not recognize the harm that will befall you. However, the combination of pressure and acceptance ("everyone is doing it") without vigilance can overcome the better nature of your ethics.

Case Histories

Unfortunately, there have been many cases of ethical misconduct in science. It is instructive to examine a small number of case histories. The first two examples were chosen primarily because they involved junior investigators. The histories are reviewed here not just to provide the facts, but also to help you sense the seeds of unethical conduct. By watching it develop in others, you may develop the inner vision you need to identify and rub it out of yourself.

Case 1: Tragic Record

John Roland Darsee was a thirty-three-year-old fellow in the lab of a prestigious professor at the Brigham and Women's Hospital at Harvard University.* By 1981, he had published over one hundred papers and abstracts while at Harvard and at Emory University[1]. This young scientist was considered one of the most remarkable young researchers working in the prestigious cardiovascular laboratory, and his seniors anticipated his becoming a leader in the treatment of heart attacks and heart failure.

However, suspicions arose over time that Dr. Darsee's work was not all that it appeared. Two of his colleagues, along with a laboratory technician who worked with Darsee, harbored suspicions about his accomplishments. Eventually, fearing that an abstract that Darsee was preparing contained no actual research but only fabricated data, they requested an investigation by the director of the lab.

The director asked Darsee to show him the raw data that was the basis of the abstract's thesis, a request to which Dr. Darsee acceded. Returning to the laboratory, he started some recordings on a single animal. However, he charted and dated these recordings in such a fashion as to make it appear that the data from this single experiment were from several experiments. Seemingly unaware of his company, he executed this fraud in the presence of several colleagues and a technician. They accosted Darsee who, in the midst of this confrontation, admitted his falsifications.

In the face of this confession, Darsee was stripped of his important National Institutes of Health fellowship. In addition, the offer of a faculty appointment was withdrawn. However, Dr. Darsee denied any other falsifications and was permitted to continue to work in the laboratory and to proceed with his publications.

During this time, the multi-institutional research study in which Darsee's lab participated was having its data analyzed. The evaluation revealed that the data collected from Harvard's lab was discordant with that of the other institutions. An ensuing investigation revealed that Darsee had falsified the Harvard contribution to the multicenter study as well.

---

* This account is taken from http://www.unmc.edu/ethics/data/darsee.htm.

The spreading investigation revealed that nine of Darsee's papers and twenty-one of his abstracts that he had authored while at Harvard were false and had to be withdrawn. Delving deeper into Darsee's past, eight of the papers that he published while at Emory were identified as false; these were retracted from the literature.

In addition, thirty-two of his authored abstracts from Emory were found to contain manipulated or fabricated data. In fact, some of the "coauthors" on several of these abstracts actually didn't know that they were listed as authors. An examination of his record while an undergraduate student at Notre Dame demonstrated that he falsified data at this preliminary level as well.

Darsee stated that although he had no recollection of falsifying data, he acknowledged the review panel's establishment of the fact of falsification and his personal role. Darsee was barred from NIH funding and sitting on advisory bodies for ten years.

To simply recount the sad story of a promising but young scientist is not a sufficient evaluation of this troubling case. The question that each junior scientist must answer is "How can I be sure that this will not happen to me?"

As Darsee came to grips with his own tragic record, he tried to explain why he has chosen to take the unethical path. At one point, he referred to the death of his father and his admiration for his mentors. At another point, he asked "forgiveness for whatever I have done wrong."

Perhaps most revealing of all, he said, "I had too much to do, too little time to do it in, and was greatly fatigued mentally and almost childlike emotionally. I had not taken a vacation, sick day, or even a day off from work for six years . . . I had put myself on a track that I hoped would allow me to have a wonderful academic job and I knew I had to work very hard for it."

Case 2: The Patchwork Mouse

A major scandal at one of the most prestigious cancer research centers erupted in the 1970s that is illustrative of the level of pressure to which junior scientists are exposed [2].

The idea of treating major human organ failure with a transplanted organ from another human was becoming an increasingly attractive idea in the midtwentieth century. In the 1960s, the concept moved from theory to reality, as surgeons mastered the monumental task of keeping the patient and the transplanted organ alive during and after surgery.

For the first time, the ability to treat end-stage renal disease with a new kidney or to ameliorate heart failure with a new heart was moving from the experimental phase to a useful therapeutic option for patients with these fatal diseases.

However, it soon became clear that most times, the body did not react well to the implanted organ. The immune system of the patient receiving the new organ responded to the transplant as it would to a foreign body, mounting a brisk and fulminant attack against the new and unrecognized tissue.

This reaction would destroy the new organ in a matter of days or weeks. This "host versus graft" reaction was not generated when the transplant occurred between close relatives or in those people who, through chance, were determined to be a "good match" with the new organ. However, for the vast majority of the population who could find no such match, organ transplantation would not be a viable option. Solution to the host-versus-graft reaction became the new imperative in transplantation research.

Attention turned to the Memorial Sloan Kettering Cancer Institute, where a young and promising scientist was brought in to solve the problem of transplant-organ rejection.

The junior scientist, William Summerlin, focused on the problem of skin grafting, where skin from one patient was grafted onto that of another. He chose this transplant procedure because the skin was an organ against which a host-versus-graft reaction would be immediately launched and easily visible. However, a solution to the skin transplantation problem would not only open the door to the defeat of the host-versus-graft reaction for other organs, but would allow skin-grafting procedures to be more readily used in unfortunate burn victims.

Dr. William Summerlin began his work by attempting to graft the black skin of one mouse onto the flesh of a white mouse. There were high expectations for his efforts, and it was hoped that he would be able to produce this patchwork mouse free of any evidence of host-versus-graft disease in short order. Finally, he claimed that he had succeeded by simply incubating the graft in a nutrient medium outside of the body, a simple solution that had apparently been missed by his contemporaries.

However, fellow researchers, not having seen his product mice, were having difficulty reproducing his results. In order to reassure his colleagues that he had indeed solved the problem, Summerlin showed his chief a collection of mice with a patchwork of black and white skin. This impressive display stayed his critics, and the transplantation community began to breathe a collective sigh of relief, believing that the host-versus-graft reaction had at long last been resolved.

Other investigators instantly queried Dr. Summerlin for his protocols to which he positively responded. Yet again, not a single investigator was able to duplicate the young researcher's findings. Finally, his superior called him in to discuss exactly how Summerlin was able to produce what no one else could reproduce.

The explanation was simple. Dr. Summerlin had simply used a black marking pen to color the skin on the white mice black.

The scientific community was flummoxed by this tragic and desperate fabrication that abruptly ended a fast-tracked career trajectory.

It has been suggested that Summerlin actually believed that he really did solve the host-versus-graft problem in the way he described; that is, using the simple nutrient bath; however, the success was tainted by the fact that he grafted the skin of one mouse to the skin of a related mouse. Unable to reproduce this successful graft, but self-assured that his results were correct and desperate to avoid failure, he resorted to the felt-tipped marker. He may have seen no harm in this because of his belief that his initial results were correct [3].

This situation was complicated, but once again, the combination of high expectations and character weakness produced a personally destructive result.

Case 3: The 2003 EL61 Controversy

Dr. Michael Brown of Caltech had been diligently tracking the object for months. Finally, checking his results repeatedly, he believed he had confirmation—the distant object in the sky, known only as 2003 EL61, appeared to be the tenth planet in the solar system.* He had been tracking it for months but told no one. Yet a competitor, Dr. Jose Luis Ortiz of the Institute of Astrophysics of Andalusia in Granada had just made a public announcement in September 2005 about his own prior discovery of the heavenly body.

Dr. Brown had planned to announce his discovery of this body in September 2005 and had issued an abstract suggesting a finding of interest in July. However, he wanted to wait for confirmation from another set of telescopes.

---

* Taken from Overbye E. One find, two astronomers: an ethical brawl. *New York Times*. September 13, 2005. D1.

Meanwhile, according to Dr. Ortiz, his own team had been scanning the heavens for objects outside the orbit of Neptune since 1992. He discovered the object on July 25, 2005, comparing it to photographs taken two years earlier in March 2003. Determining that 2003 EL61 was a new planet, Dr. Ortiz shared the information with Brian Marsden of the Harvard-Smithsonian Center for Astrophysics, director of the International Astronomical Unions Minor Planet Center that served as the clearinghouse for these discoveries.

After validating the observation for themselves, Dr. Marsden's team disseminated the details to the rest of the astronomical world on July 27, 2005. Thus, Dr. Michael Brown's plan to make his own announcement at a meeting in early September 2005 was trumped by Dr. Marden's announcement of Dr. Ortiz's success. Dr. Ortiz's accomplishment was a surprise, however, because his telescopes were relatively unknown.

However, a detailed assessment revealed a more nefarious turn of events. Dr. Brown posted his own observations of this astronomical body on the internet in July 2005. Investigation revealed that after this posting, someone from Dr. Ortiz's lab did an internet search for the data mentioned in Dr. Brown's abstract and was linked to the electronic observation logs of Dr. Brown. Thus, an internet search would take the user straight to the observations on which Dr. Brown based his abstract.

A check of internet protocol (IP) addresses revealed that someone from Dr. Ortiz's lab accessed Dr. Brown's observation logs eight times between July 26 and July 28. Each time, the intruding computers went to the specific part of the site devoted to observations on the astronomical body. Furthermore, it was determined that the IP address of one of the intruder computers was the same as that of a computer that sent emails from Dr. Ortiz to Dr. Marsden, announcing the "discovery." Brown's group accused Ortiz's group of a serious breach of scientific ethics and asked the Minor Planet Center to strip them of discovery status.

Dr. Ortiz criticized Dr. Brown for being too slow in announcing his discovery and for "hiding objects." He further said, "The only reason why we are now exchanging email is because you did not report your object." Ortiz later admitted he accessed the internet telescope logs, downloading the relevant information a day before making his announcement, but denied any wrongdoing. In the end, he conceded that it was Brown's team that had discovered the object.

Fraud and Experienced Researchers

According to the National Science Foundation, there were 50 cases of misconduct involving research sponsored by the National Science Foundation and 137 cases involving research financed by the National Institutes of Health in the decade ending in 2002. Fraud can reach the most prestigious of scientific centers.

Dr. Luk Van Parijs, thirty-five years old, a respected scientist in the use of ribonucleic acid in studying disease mechanisms, had published articles in influential journals, for example, *Science, Nature Genetics*, and the *Proceedings of the National Academy of Sciences*. A researcher at the Massachusetts Institute of Technology, he was a respected immunology researcher at its Center for Cancer Research.

After the publication of a manuscript, members of Dr. Van Parijs's team came forward with accusations of misconduct. Dr. Van Parijs was placed on leave of absence while the investigation took place.

The investigation confirmed that fraud had taken place and that it affected one published paper and several articles under consideration for publication in the peer-reviewed literature. Dr. Van Parij admitted the fraud, and the report was sent to the Office of Research Integrity in the Department of Health and Human Services. Upon confirmation of the fraud, Dr. Van Parijs was dismissed from MIT.

"I Promise"

"Can you stand up and make me walk again?" the son of the minister asked the scientist standing above him commanding the room's attention.

Leaning over, Dr. Hwang Woo Suk replied, "You'll walk again, I promise."

It was 2004, and Dr. Hwang was riding the crest of national enthusiasm for his own work.* From 2003 to 2005, Dr. Hwang became a hero of South Korean science, claiming a series of fantastic biological breakthroughs. By the end of 2005, his work had been debunked, his results revealed as fabrications.

One of the most controversial and competitive areas of modern biology investigation in the early twenty-first century is stem cell research. However, the area is fraught with technological obstacles, a major one of which was the process of cloning adult stem cells. This

---

* Taken from (Sang-Hun Choe, Wade, Nicholas, December 24, 2005. Korean Cloning Scientist Quits Over Report He Faked Research. The New York Times. A1.)

procedure, a required early step in the development of organ regeneration and replacement, was an exceedingly slow and wasteful one.

In 2004, Dr. Hwang announced that he had successfully cloned a human cell by inserting an adult cell's nucleus into a human egg. Of equal importance was the efficiency of the new process. While it was not unusual to use 242 eggs to establish a cloned human cell line, Dr. Hwang claimed to reduce this to 11. In June 2005, his office claimed that he had repeated this process for eleven patients. Following on the heels of this assertion was the claim that he had cloned a dog as well.

Disturbing questions emerged in the fall of 2005 when reports appeared from the lab of Dr. Roh Sung II, asserting that women who worked in Dr. Hwang's lab may have donated eggs for an experiment in cultivating stem cells from cloned human embryos.

When news of this revelation was made public, Dr. Gerald Schatten of the University of Pittsburg severed his research ties with the South Korean research team because of ethical violations. In addition, Dr. Schatten, writing to the journal *Science*, asked that his name be withdrawn as senior author from one of the articles submitted with Dr. Hwang. Alleging doubts about the manuscript's accuracy, he now believed that some of the experiments may have been fabricated.

Hints about the existence of these discrepancies came from a small group of young South Korean scientists who posted disturbing contradictions of Dr. Hwang's work on the web in the form of photographs.

The one that appeared in the June 2005 *Science* article about the eleven patients also appeared in the October nineteenth issue of the *Biology of Reproduction*, which stated the photograph was one of standard stem cell colonies.

Thus, the reported results from all eleven patients represented not a breakthrough, but merely copied photos of the stem cells produced from surplus embryos. In the minds of his critics, this falsified photograph cemented the argument that Dr. Hwang had never successfully cloned any adult human cells. The meltdown of this promising line of research was particularly surprising to many visiting scientists who were impressed with the technical skills of his lab and the devotion of its workers.

Dr. Laurie Zoloth, director of the Center for Bioethics, Science, and Society at Northwestern University, stated, "We depend entirely on the truthfulness of the scientific community. We must believe that what they are showing us and what they say has been demonstrated is worthy of our concern and attention." The deep desire for dramatic advances in a promising field in combination with a well-established trust of contributing scientists was shockingly betrayed.

Human Together

As you read these retrospectives, it is easy to react in astonishment to the poor judgment of the protagonists. However, the question is not whether you agree that their actions were lamentable but whether you're in danger of succumbing to the same combination of external pressures and heartfelt desires that, along with exhaustion, led to the downfall of these scientists. There are circumstances that will produce unethical behavior from everyone. To which are you susceptible?

For example, we believe that we are honest, but how honest are you when you're exhausted, under great personal strain, and are the subject of abuse by your overbearing superior? Is there no point where your integrity will break down?

Devote your energies to repairing the vulnerabilities of these situations by (1) avoiding the environment that reveals your weakness until you can build your emotional strength and character and (2) change your orientation from a focus on the needs of your weakness to the high standards required by your acceptance of your growing character and own great value.

This tack might require you to shun the high-pressure rapid career advancement track if the strain that it produces allows your weaknesses to emerge.

Until your character grows and you can maintain steadfast honesty in the face of fatigue and pressure, choose to be content in a different position in which the best of your talents can be brought to bear and displayed rather than work in a more prestigious position that will bring out the worst in you, destroying your promising career. Don't progress as fast as possible but instead as fast as is consistent with your ability to work reliably, dependably, and honestly.

> **Don't progress as fast as possible but instead as fast as is consistent with your ability to work reliably, dependably, and honestly.**

Additionally (when rested), examine yourself. Force to the surface the deeper issues that command your unethical behavior. What is it that you fear will happen if you are honest and admit a failure of an experiment or research program? Why are you so frightened of failure that you're driven to considering data falsification?

> **There are circumstances that will produce unethical behavior in everybody. Be on guard for your vulnerable environments.**

If the fear is self-condemnation, then you have a great deal of company. It is commonly not the failure at a task that we fear, but the accompanying self-loathing that we cannot bear.

The simple but revealing truth is that you, with your intelligence, insight, and capabilities, are far more valuable than your experiment's outcome. Begin to replace the self-condemnation with self-respect. Sometimes this requires counseling while at other times, simple but continued self-reflection is sufficient.

Also keep in mind that if you do not have the particular weaknesses revealed in the previous case histories, then you have others. We are all human together. We do not each have the same flaw, but we each have some flaws against which we must struggle. Come to know your particular set of weaknesses and work to strengthen yourself against them.

Pervasive

The preceding egregious examples received attention in both the lay and scientific communities. However, the environment in which these glaring examples took root is all too fertile for the development of scientific misconduct.

In a commentary, Martinson et. al. [4] describe the results of a survey they sent to several thousand scientists funded by the National Institutes of Health asking these researchers to report on their own scientific conduct.

The 3,247 anonymous replies these investigators received revealed a wide range of questionable practices. The proportion of reports that admitted serious conduct infractions was low; 0.3 percent reported fabricating research data, and 1.4 percent admitted to plagiarism. However, other lesser offenses were more prevalent. For example, 4 percent of responding researchers admitted to publishing the same data in two or more publications to beef up their résumés. Just over 13 percent reported using research designs they knew would not give accurate results.

Slightly over 5 percent of respondents reported that they threw out data because it contradicted their previous work or admitted that they had circumvented human-subject protection. Approximately 10 percent stated they had inappropriately included their name or those of others on published research results. In addition, 15 percent stated that they changed a study design or results to satisfy a sponsor or ignored observations on a hunch that they were inaccurate. These proportions were fairly uniform across the spectrum of experience, from junior scientists to their more senior colleagues.

While the survey suggests many possible explanations for this misconduct, including the frustration researchers feel for what is widely perceived as an inequitable reward system, one concern expressed by the authors is the new scientific era of competitiveness.

Martinson stated, "We've turned science into a big business but failed to note that some of the rules of science don't fit well with that model." While the most prevalent misconduct is the least serious, its frequency speaks to the high-pressure environment in which researchers work. Given the relatively high prevalence of minor misconduct, it comes as no surprise that extreme egregious behavior is occasionally generated, its conflagration springing up from ever-present background brushfires.

Ready, Shoot, Aim

One of the most pervasive influences of the current scientific environment is the sense of urgency. Scientists work under the continued requirement of productivity. With the large number of meetings, abstract deadlines are omnipresent.

Manuscripts currently in the peer-reviewed process commonly require a rapid response to the reviews to maintain their position in the publication queue, scientists working in the private sector labor to meet difficult production schedules. While the drive to be productive is important, a casualty of compulsive productivity is careful thought and reflection. This rapid-fire approach to generating results undermines the one unique natural talent scientists have: careful, detailed, reflective thought.

If we define science as the intellectual and objective pursuit of the truth, then a corollary of this definition is that this pursuit must not be a hasty one. Science is not about speed. Science is about accuracy. It is about precision. It is about the careful and thoughtful review to root out mistakes in reasoning and judgment.

Other worthy endeavors can be executed quickly. Capitalism is about speed. Bonuses are about speed. Promotion and tenure are about speed. Profit is about speed. Not science.

It is too much to ask that a lone scientist change the system. But a single scientist can change themselves. This change is an important one. You have been provided the talent. Insist on the right to apply that talent to its fullest.

Generate the environment that allows you to do your best work, acknowledging that this will hurt your productivity. Miss an abstract deadline if you have not checked your results for accuracy. Shun a brutal production schedule because it precludes you from doing your best work.

These losses in productivity can be serious. You will be criticized for "putting a monkey wrench in the works." Your career trajectory may be altered.

There is no way around this observation. Choosing to create, work in, and fight for an environment that permits you to conduct ethical research will cost you. It takes courage and the prior reflection that your career trajectory may suffer. Character development and protection help you to endure this sometimes grievous setback. The justification for it is the lifelong dividend that comes from the strength of character produced from this self-investment and high-quality work.

> **Create, work in, and fight for an environment that permits you to conduct ethical research.**

Careful Citations

With the intense competition for grants, publications in prestigious journals, and even media coverage, it is all too easy to view career progress as a zero-sum game; a tournament in which every victory achieved by one research group is felt as a defeat by another.

This is a remarkably prevalent mindset in endeavors that are, by definition, collaborative. While very few motivations fuel our productive efforts like healthy competition, rarely does the solo scientist working alone and in isolation miraculously produce a stupendous result. However, competition remains only a device; the nature of our work remains collaborative.

As a junior investigator, there are two clear areas of your work in which you'll have the opportunity to put your ethics on this issue to the test. The first is your ability to resist taking credit for a capable student's good idea. The second is your reaction to the good idea of a colleague that you come across in a manuscript review or grant-application review.

The common thread that connects these situations is that the originator of the idea is in a disadvantageous position. The naïve student cannot adequately protect herself from intellectual property theft, nor can the manuscript or grant author defend himself from the malevolence of an anonymous reviewer. Each circumstance provides the opportunity for theft, precisely because the victim's work product is open and exposed.

Specifically, it requires ethical strength to resist the temptation of stealing from a vulnerable colleague.

Deft Thefts

The work of students is especially vulnerable to intellectual theft. Students develop freely in a trusting intellectual environment that encourages the flow and interchange of ideas. By being assigned to a student as their advisor, mentor, or thesis/dissertation committee, you'll be exposed to many good and innovative ideas. Commonly, very little thought is given to linking or tagging each idea to its correct creator in this intellectual cauldron.

Several of these ideas may be important and useful to you and, when properly developed and described, would make a worthy contribution to your field. It therefore should come as no surprise that the temptation can be great to take the student's idea as your own. The student may not recognize this deft theft as such and, in their naiveté, may believe that your action is appropriate.

However, taking the student's ideas and work product as if they were your own is naked intellectual theft. Like any crime, several people are damaged in the process. One is the student (and indirectly the student's family); the second is you and your character.

Commonly, the student's goal is not just to find a topic, but to gain the widest possible experience in developing that topic. They are looking for new and formative experience with cerebration, instrumentation, computation, but most importantly, collaboration.

They have worked hard and patiently to put a committee of advisors together in order to produce the highest caliber of scientific and professional interchange. They assiduously pursue every lead that you provide for them and excitedly share with the committee any new ideas or breakthroughs that they believe they have experienced.

They are anxious to make progress, but that progress is built on a bedrock of trust and loyalty that develops between the committee and the student. The student draws freely on that trust, using it to shape their scientific thought and judgment.

Intellectual theft demolishes that foundation of trust. The student's work productivity is thwarted by the theft. Additionally, and perhaps most importantly, the student herself is damaged by the encounter. Her ability to work effectively with others is impaired, and only with great difficulty is she able to trust other collaborators again.

Overcome the temptation to usurp the student's work as your own. Avoid the trap of thinking that you'll do a better job than the student in placing the student's work product in its best context and in its promulgation.

Who can do the better job is not the point; the student has earned the right to complete the development herself. The value of her work product is secondary to the value of the experience she will gain in taking the lead in publishing and disseminating the result.

If you're concerned about a successful integration of her work into the current fund of knowledge, then the best role that you can play is a supporting one, letting her draw on your expertise in crafting her report or publication.

The student benefits immensely from this concerted effort on your part and, in the process, gains a remarkable experience in seeing what a good professional collaborative writing experience can produce.

A second thought trap that investigators fall into is the belief that they need the acknowledgment of the work product more than the student does. The investigator may think that the student has a wealth of opportunity in front of them while they do not.

This is a remarkably pessimistic and self-destructive thought process. Both you and the student are better off if you devote your support efforts to helping the student publish their work in their own right, with them as first author and you as supporting author. This can be a seminal experience for a student researcher while simultaneously rejuvenating your career interest.

Grant and Manuscript Review Theft

Grant review participation can be both overwhelming and rewarding. At the forefront of research, the conversations concerning the grant's scientific content illuminates research imperatives while displaying new and exciting methodology. The literature reviews are commonly excellent, and the interchanges that take place among the reviewers are vivid and provocative. Involvement in grant reviews is a wonderfully productive activity for junior investigators, providing the opportunity to both learn, discuss, and debate contemporary issues that surround your research interest.

However, reviewing a grant is essentially looking closely into the ideas and intellectual repository of a fellow investigator. When the work of others is open and unprotected, the temptation of theft is near at hand. Unscrupulous grant reviewers can pirate an idea, a technique, or a style of analysis from the grantee whose work they are reviewing, taking the idea as their own.

This theft debases the reviewer, the grantee, and the entire review process. When the grantee discovers that his work has appeared in an unacknowledged form, a dispute can break out, further eroding scientific relationships and wasting valuable time.

This is an unfortunate sequence of events, made all the more so because the coveted ideas will commonly become available later in published form. If the reviewer finds themselves unable to wait, they can call the grantee and discuss the issue after the grant review has been completed (including formal notification of the scores the grantee received on the review).

This kind of postgrant communication can be very productive and lead to a new collegial relationship between the reviewer and the grantee. By exerting discipline, the grant reviewer has elevated the work of the grantee and the nature of his relationship with the grantee.

Similarly, when the peer review process for a manuscript is complete and the reviewer has received the final decision of the journal to which the manuscript has been submitted, it is quite appropriate for the manuscript reviewer to contact the author to discuss the idea suggested by the author that intrigued the reviewer. Issues of acknowledgment should be covered to the author's satisfaction. Once this is accomplished, new collaborative efforts can be professionally and ethically forged.

Cite Honestly

Modern problems that scientists are compelled to solve are by nature constructed so that no one working singly and in isolation can solve them. Instead, the sustained, combined efforts of different individuals with different talents is required to attain the final solution.

It is sometimes difficult to see where one scientist's contribution ends and the other begins within these collaborative constructs. Which of two men who are both struggling to move a heavy piano up a flight of stairs can take credit for completing the task at the expense of the other? The burden of labor may shift momentarily from one worker to the other, but the joint effort of each is required. In the end, the feat is accomplished, and both share in the achievement.

Since your work product's luster has an important contribution from the reflected light of the accomplishments of your colleagues, learn to take a singular delight in acknowledging the work of others.

Be careful and specific in detailing the contribution that other scientists have made to your work in your own writings and presentations. If you make a mistake in a citation, move rapidly to correct it, as you would hope that others would react to a mistake in the citation of your own work. Work openly and collaboratively, freely acknowledging the work of others.

Finally, those who attempt to usurp what in reality belong to another can be in for quite a surprise, as demonstrated below:

> When a man attempted to siphon gasoline from a motor home parked on a Seattle street, he got much more than he bargained for. Police arrived at the scene to find an ill man curled up next to a motor home near spilled sewage. A police spokesman said that the man admitted to trying to steal gasoline and plugged his siphon hose into the motor home's sewage tank by mistake. The owner of the vehicle declined to press charges, saying that it was the best laugh he'd ever had.[*]

Being less than honest can produce its own unpredictable collection of unfortunate surprises.

---

[*] Taken from the Darwin Awards 2003, which may be located at http://www. binarywrangler.com/archives/000063.php.

Spinning Results

Your work must be promulgated if it is to be received and integrated into the current fund of knowledge by the scientific community. Ultimately, you will have the primary, if not exclusive responsibility in this affair and therefore bear a special and sometimes solitary burden.

Unfortunately, research results are frequently described in a mawkish rather than circumspect manner. This is understandable. There are only a small number of presentations that can be made at highly visible plenary sessions occurring during prestigious national and international meetings.

Similarly, almost every scientific field has a limited number of top-quality journals, producing a natural and healthy competition for publication slots. In order to gain the widest readership (and perhaps the greatest notoriety), the temptation is great to state the research results in their most provocative manner.

These overstatements are misleading and reprehensible. This sensationalism is further amplified by the lay media.

The price science pays for this tabloid-type journalism is inaccuracy and, ultimately, loss of credibility. The greater the stretch from reality, the greater the potential to mislead the community. It is clearly critical to describe the implications of your work product. However, it is also important to describe these results cautiously.

There are two useful approaches that you as a junior investigator can take to purge your work product of this detritus.

The first is to affirm that you'll be satisfied with nothing but your best scientific effort. As a young scientist, you have developed and demonstrated scientific intuition and capability. It takes practice and consistent, patient effort to wield these in the right combinations.

Now, develop the internal discipline and strength that you need to suppress your work if it is not the product of your best effort. Only when you have satisfied yourself that you have before you the result of your finest work product should you disseminate it, allowing it to stand on its own merits.

Secondly, consider who your research effort will serve. Is its purpose merely to increase the citations in your own résumé? To speed you along your career trajectory? When you write, let your writing be motivated by the need to serve. When you make a contribution to the literature, let it come from spirit of service, not of the need to feed your own curriculum vitae.

Service, not self-aggrandizement, is central to making a solid contribution to science.

How will the results of your work best serve the readers? An honest answer to this question will purge any hint of intellectual dishonesty from your papers. It also creates the intellectual atmosphere for you to write freely about the limitations of your work that will help your colleagues best determine what the implications of your findings truly are.

Conclusions

Ethics is an approach to life and not a mere collection of rules. Your ethical behavior is the living expression of your core principles that govern your relationships with people. Just as you have self-worth on which you rely, your sense of the worth of others regardless of their opinions and actions governs your ethical treatment of them. Institutional rules help you operationalize your beliefs; they do not replace them.

Ethical behavior is dynamic. Therefore, as your field evolves, so too should your behavior toward others improve. If it is your intent, as a junior investigator, to be an ethical scientist, then challenge, reinspect, and if necessary, readjust your conduct.

A scientific career that starts ethically will in all likelihood not end ethically if that scientist does not stay current with ethical issues and the impact of these issues on her field. If your central belief is to treat people and animals with respect, then small adjustments in your conduct that a developing scientific field requires should be easy to make.

Most importantly, remember that ethical researchers are not perfect. Ethical people make honest mistakes. However, what characterizes the ethical scientist is the response to their mistake. When they recognize that they missed an opportunity for ethical conduct, they apologize, make appropriate restitution, and having learned the right lesion from their error, they move on. Ethical behavior is not perfect behavior. It is behavior that calibrates and self-corrects.

Insist on working with other ethical scientists. Do not succumb to the belief that you're justified in working with an unethical individual because of the productivity boost that your career will get. This is like scooping scalding coals into your lap in the hope that you can stay warm; in all likelihood, you'll be burned in the matter.

> **Ethical people make honest mistakes. However, what characterizes the ethical scientist is her response to that mistake.**

Also, when an ethical infraction takes place and your "ethical alarm" goes off, do not treat it like the alarm that wakes you in the morning, hitting the snooze button and then going back to sleep. Treat your ethical alarm like a fire alarm.

You have invested time and effort in developing and clarifying your character and ethical beliefs and sharpening your vigilance; do not ignore them when they attempt to warn you of an urgent problem. After recognizing the difficulty, think carefully before you speak on ethical matters. This is perhaps more important here than in other circumstances because the words and their implications can give offense where you don't want it to. In all matters, but in this matter especially, know what you're going to say before you say it.

Finally, good ethical decisions require good judgment, and good judgment is most easily found when you take care of yourself. You cannot bring the best of yourself to an important decision when your body is starved of sleep, your mind is starved of rest and relaxation, and your stomach is starved of food. Make these decisions when you're rested, well-balanced, sharp-eyed, secure, and confident.

References

1   http://www.unmc.edu/ethics/data/darsee.htm.

2   Hixson, J. *The Patchwork Mouse. The Strange Case of the Spotted Mice.* Anchor Press.

3   Medawar PB. (1976).*The Strange Case of the Spotted Mice.* The New York Review of Books; 23. http://www.nybooks.com/articles/article-preview.

4   Martinson BC, Anderson MS, deVries R. Commentary: scientists behaving badly *Nature* **435**:737-738.

17

Sexual Harassment

There is a wealth of material on sexual harassment, and the advice that is so readily available elsewhere will not be repeated here.* There is no excuse, and no person (junior scientist or otherwise) should ever tolerate or inflict sexual harassment or abuse.

Make no mistake. Your first obligation to your colleagues is to develop and sustain a professional relationship with them. This involves treating them, their opinions, their work, their personalities, their emotions, and their bodies with dignity and respect.

This is an unceasing, relentless obligation that you must face and discharge regardless of whether you have the desire and your colleague's permission to explore a nonprofessional relationship. If you do pursue a personal relationship with a colleague, then that personal relationship does not replace your professional obligation to them. Any personal aspects of your relationship must be in addition to your responsibility to treat them professionally.

The complications of emotional and sexual involvement can be overwhelming, and only a very few people are able to develop and sustain

---

* See, for example, http://www.eeoc.gov/facts/fs-sex.html. A statement of the federal law that governs this is http://www.eeoc.gov/policy/vii.html, and an interesting discussion of the complexity of the topic may be found at http://www.menweb. org/throop/harass/commentary/hostile-env.html.

them while maintaining the true interpersonal dignity that professional relationships demand and require.

It is rare for a couple to work together as vibrant professionals yet develop and retain a personal and intimate relationship, an observation that acknowledges the particular emotional balance these couples must have. The rest of us are better off avoiding the entire matter by keeping our emotional and sexual interests separate and apart from our professional relationships.

Redline

Like ethical issues, sexual relations between colleagues are complex. Clarify this complexity by drawing your "redline," that is, behavior in a colleague toward you that is unacceptable.

This is a natural product of your character sabbatical. Doing these reflections, you have considered and discarded many redlines until you have found one that fits. Once you have it, own it and use it, accepting no breaches. If your redline includes shoulder squeezing, then remove their hands and say clearly, even loudly, "Do not do that ever again."

Don't worry about making a scene. Your colleague made the scene, not you. You simply ended it.

While there may be times to be bashful, this is not one of them. This is self-defense. Be clear, plain, and assertive. You are not causing the problem—they are. If they insist, then overcome the resistance. Leave the room, ask someone for help, and then document the event, filing a complaint with your human resources department.

And if you work in a predatory environment with no good recourse, simply leave. Retire. Quit. There is simply no reason good enough to stay in a personal environment such as this for you to devalue yourself by staying.

Reassert your value by leaving.

18

Ethics and Difficult Superiors

The detection of unethical conduct requires vigilance, and vigilance requires a metric.

While the presence of unethical behavior in a senior investigator can be flagrant, the suspicion arises slowly and can be disorienting. To the junior investigator, the project's senior scientist is a person who has made the impact on the scientific community that the junior researcher hopes she will be able to make. Sometimes, the senior investigator can be a hero to junior researchers, leading to the junior deferring to the senior in questionable circumstances.

A requisite for the junior investigator, however deferential they might be to their senior scientist, is the development of her own ethic. This evolution has its foundation in her own character, culture, and family, but is further refined based on the interactions the junior investigator has with others through discussion, experience, and reading. The development of this professional ethic is best nurtured actively and consciously rather than unconsciously. It requires good input from independent sources, as well as reflective thought.

Just as your technical knowledge will never cease growing, so too your character and ethical development should not stop. Challenge yourself. Can you articulate your ethical point of view? Can you describe the fundamental principles on which it is based? Can you calmly defend your ethical positions?

In science, we quickly learn how to form arguments in order to defend our technical perspective. Typically, we are not as able to describe and defend our ethic. While this may have been acceptable in the past, the time for ethically mute scientists has come and gone.

It is most helpful if this produces an informed discussion between you and your colleagues. The point of this discussion is to provide you an opportunity to examine your ethical perspective from another point of view. Since perspectives in science and sociology are not static, it comes as no surprise that ethical values and standards must evolve.

Trust, but Verify

By working with a strong-willed chief who has far more experience and expertise than you have, it is natural for you to give this chief the benefit of the doubt in technical and scientific matters. However, this natural deference can make it all too easy to replace your ethical sense of direction with that of your chief's. While trust in the ethical judgment of your chief may be appropriate, you must also challenge your senior scientist on ethical issues in which you believe you may have a different point of view. Therefore, ask and discuss the ethics of the troubling situation with your chief.

It is critical that you both retain and develop your own sense of ethics. Do not engage in a wholesale replacement of your ethic with that of your chief's sense of values. Of course, you'll learn from the senior scientist if it is appropriate to choose to allow his experiences to shape your sense of propriety; however, this process requires you to understand and affirmatively assimilate portions of his ethical sense of direction into your own.

This is distinct from the wholesale embrace of his ethical values that is permitted to proceed because "he's the boss" and therefore must know best. The first is a careful assimilation process where you discard parts of his ethic that you believe may be wrong. It requires thoughtful consideration and careful integration. In the second, you accept all that is his, simply because it is his.

Fortunately, most chiefs are willing to spend some time talking about the ethics of a situation. Make it clear that you would like an explanation to help further calibrate your ethical compass. It is more likely than not that the senior scientist's ethical perspective is better balanced than yours, or alternatively, that you're incompletely informed.

For example, there may be some history behind the particular situation of which you're not aware but in which your chief took part. These past events can add a completely new dimension to the ethical issue under discussion.

If on the other hand your chief is unwilling to discuss these issues of ethics with you, then frankly, you have a new problem, and one we will discuss momentarily.

Delegation

A well-accepted and admirable trait in a senior investigator is the willingness to delegate responsibility. In addition, junior investigators looking for advancement opportunities gain valuable experience in carrying out new activities under the tutelage of their superior.

In the overwhelming majority of circumstances, this relationship provides important benefits to not just the investigators, but for the project as well.

However, a difficulty can arise when the senior investigator asks that his junior take a step in the discharge of the delegated task with which the junior investigator ethically disagrees.

For example, a statistician working on a research project in a competitive research project may be pressured to produce "a significant result" and commanded to "get the significant result any way that you can." An investigator, interested in gaining some skill in reviewing budgets, is told to "reduce the budget for the lab to $60,000 per year, and I don't care what it takes." A research physician may tell his resident, "I don't care how you do it, but I want that patient in my study."

It is best in these circumstances for junior investigators to do two things. The first is to recognize that their actions first and foremost must be governed by their individual sense of ethics. It is no longer acceptable for you to carry out an unethical request, a request that you know to be wrong, simply because your boss asked you to do it.

When a final accounting for the event occurs, you will have to explain your actions. Let the motivation be something other than "I knew as a professional that the action I was asked to take was unethical, but I did it anyway because my chief told me to do it."

Let your character command here.

In the end, if the action that you have been asked to take does not meet your own ethical standard, then don't carry out the action and state clearly why you're refusing. This stand on your part will only rarely be demanded, but be assured that at some time, it will be required. This requires courage because your career trajectory may very well be damaged.

Secondly, junior investigators have to learn to hear things the way they are meant to be heard. When a superior investigator says to carry out an activity and "do it any way that you can," he commonly does not mean this literally.

There is no doubt that this is what he said, but it is not what he meant to convey. What he meant to say was "Do this in any way that is practicable and ethical." However, the press of time and his own frustration truncated the message.

If you have a doubt about this, then challenge him by saying, "You mean, 'Do this in any way that is practicable and ethical,' right?" This may lead to a moment of embarrassment, but far better to experience the moment than to stay silent and find yourself on the horns on an ethical conundrum. A solid character permits you to challenge your senior in this matter.

Final Comments

Familiarize yourself with the regulations that govern the definitions and guidelines of ethical activities at your institution. They serve by setting the appropriate standard for ethics. However, also recognize that trying to write rules that govern all matters of ethics is like trying to count the grains of sand on a beach; while many will be enumerated, many more will be missed.

These will be helpful for the next chapter.

19

Vexing Bosses

Among the most vexing problems that junior investigators must confront is the problem of interpersonal difficulties between you and a chief scientist to whom you report. Successfully navigating these difficult waters requires self-discipline, balance, and superb interpersonal skills. In these disputes, your goal should be to address the difficulty sensibly and professionally. Choose a behavioral level that allows you to emerge from the situation with a stronger disposition and sense of balance. You cannot guarantee a satisfactory outcome to the dispute; however, you must ensure that your efforts to resolve this matter reflect your character and personality.

Castigation

Out of the blue, you get a phone call from the administrative assistant of the program's senior investigator. Although the language is polite, the tone is unmistakably clear. The principal investigator of the project needs to see you at once.

The reaction of the junior investigator is immediate and common. Hearts beat fast. Palms sweat. Stomachs churn. If the meeting is not to take place for a day or so, you may have some restless nights in front of you.

Since these types of meetings will punctuate your career, you'll need some usefully coping skills to help you prepare for and get through them.

One of the first skills that you require is one that effectively deals with the issue of intimidation.

The Intimidation Factor

It is not your superior but instead your fearful reaction to them that is your greatest adversary in the upcoming meeting.

Take this enemy head-on by first recognizing what intimidation does to you. The internal pressure and fear generated by intimidation disconnect you from the best of your abilities. Specifically, fear steals your memory for details and your ability to calculate. Trepidation causes your fine intuition to escape you and can be so bad that you lose your ability for articulate comment.

Of course, these faculties are still present, but the anxiety has built a wall up between you and your talents and capabilities.

In a very real sense, fear has caused you to disarm yourself when you need these abilities more than ever. Letting fear into your heart and mind is like letting a thief into your house. You're going to be robbed. Worse, you're complicit in the theft.

To retain command of your own faculties, you must demolish your fearful reaction. Begin by recognizing that fear (like fungus) grows best when it is left alone in the dark. Deny fear its strength and hold over you by pulling it out of the shadows and into the light.

It will tell you that the upcoming meeting holds the fear of criticism, demotion—judgment.

Thus begins the cascade of failure. You fear being ridiculed. You fear the loss of status, the denial of a recommendation in your behalf, the fear of job loss, the fear of an uncertain career, a damaged life, self-repudiation.

However, these fear-melded chains that ensnare you are made not of iron but of vapor.

The reality is that even without facing your superior's wrath, your career is uncertain.

In fact, you can't even guarantee that you'll have a safe drive to work tomorrow. If that can't be assured, how can your future career, affected by forces that you cannot see and may never know, be certain? Your character development helps with this.

The reality is that your future will always be unknown, regardless of how your boss feels about you at the moment.

However, there is a second balancing reality. It is that uncertain futures are no threat to those with self-worth—the endowed.

Capable people develop the strength and spirit that they will require to deal with the stresses and challenges in the future. Specifically, the gift to talented people is that if they are willing to look, they can find within themselves a new combination of strength, insight, discipline, vision, and energy to confront future problems. You can therefore be assured that for the unseen problems of the future, you'll have new and heretofore unseen strength. Being therefore assured that the future will be handled by these abilities fully energized when the need for them arises, you can be confident in the face of future uncertainty. This confidence for the future releases you to focus today's energies and abilities on today's problems.

Thus, the way to defeat fear is to simply see that fear is based on a lie. Destroy it with the simple truth and assurance that your developing natural talents and abilities can overcome uncertain problems waiting in your future. You're more than a match for whatever will face you. No senior investigator or boss, no matter how demanding or intimidating, can take this realization from you. You may rob yourself of it by giving yourself over to your fear, but no one else can rob you of this truth. Only you can do that.

Put another way, anything that your superior can take from you is something that in the long run is not worth having. Accepting this truth when you walk into the room to meet your boss can profoundly affect your discussions with them.

Listen

Once intimidation and your fearful reaction have been rubbed out, you're free to bring your best faculties and talents to bear in the meeting. One of the most important of those is the ability to listen. It is much easier to fully listen in the absence of fear. Listen carefully to what your senior investigator is telling you. Be sure to ask questions to understand exactly what she's conveying. Try to appreciate her point of view.

This is particularly useful if you are being criticized. Even though you may disagree with what she has to say, make sure that you understand the position that she is taking. As a senior investigator, she has more experience than you do, so allow yourself to give full consideration to the possibility that she might be right.

With your sense of value being unaffected by the discussions, you can afford to slowly and carefully consider her comments. If a response is not required right away, take a day or so to consider exactly what you have heard. Keep foremost in your mind that a lecture from your senior investigator is, above all, an opportunity for you to learn. Like a short-term loss on a smart long-term investment, be willing to accept the occasional small measured steps backward on a good, well-planned career path.

An occasional corrective lecture, while not enjoyable, can propel your character development forward. The events leading to the meeting may reflect a weakness on your part, a weakness that has now been forced into the open for you to critically review and examine. Evaluate this flaw carefully, taking deliberative steps to (1) strengthen this weak area and (2) be vigilant for the circumstances in which this failing may be exposed again.

This second point is critical. While you quite naturally may want to avoid the circumstances that revealed your weakness, you nevertheless will have to confront it repeatedly during your career. Learn to anticipate these circumstances, overtly considering what your correct and appropriate response should be. Since this is a weak area for you, its repair and restoration require your careful consideration and attention.

Delivering a Message

Principal investigators, chiefs, and department heads have many different motivations for being harsh with particular junior scientists; there are chiefs who are difficult, chiefs who are personally offensive, and chiefs who are unethical. What these senior investigators all have in common is that they each create a contentious work atmosphere. However, you must differentiate between these different personalities and styles because while working with the difficult, tough boss can serve you well, working with the personally offensive chief or the unethical senior investigator can damage both you and your career.

The tough and unfair boss consciously singles you out for special critical attention. He tends to be short on praise, but provides extensive critical discourses and elaborative disapproval. Despite your best efforts, he is unsatisfied and impatient, tersely demanding to know why you haven't taken the next step in your work. He seems to go out of his way to put you in situations (e.g., making presentations to audiences) in which he knows that you'll be uncomfortable.

Central to your appropriate reaction to a hypercritical chief is your careful self-evaluation. It is common to wonder if you were misinformed about what your role would be in the project. Perhaps there was a misunderstanding about the talents that you would be bringing to the team. If you can, speak to other colleagues who have trained under him to see if they have been the object of similar treatment.

In this situation, it is very easy to give yourself over to despair because he may have damaged your feelings. However, it is also possible that your chief has something valuable in mind for you. Possibly, his actions were calculated not to produce the best environment for you, but instead to create an atmosphere in which he commands your full attention. He has something to say to you and would like your full concentration (perhaps in a way that you have never concentrated on anything in the past) before he can say it.

This message is a core message that can be central to your career development. It may be that he has identified a penchant for the self-indulgent in you or that he believes that you pay insufficient attention to your instrumentation. However, whatever his perceived shortcomings of you, he believes that they are dangerous enough to damage your career trajectory. The persistent criticisms that you received from him so far are like the light but steady rain announcing the coming storm that while tempestuous, it provides the necessary water for good growth.

Now, of course, it would be better for all if your boss could just come to talk to you about this issue openly, honestly, and directly without this complicated behavioral preamble. However, interpersonal relationships and, frequently intergenerational relationships, are complicated by expectations and personality nuances. A boss who cared nothing about your character, scientific progress, or career would not take the time to identify your problems and share his perceptions with you. In fact, your chief recognizes your talents and would like to see you make the fine adjustments necessary for your continued productive work. Since this type of chief has your best interest at heart, he is most likely to respond to your reaction to the environment in which he insists you work.

If this is the case, your task as junior investigator is to prepare yourself to hear this message. This task begins with ensuring that you have a solid and reliable sense of self-worth and value, separate and apart from your external critical environment. This strong sense blocks the conversion of harsh and helpful criticism of you into destructive self-hatred, as discussed in earlier chapters.

In preparing to meet with him, focus on two truths about yourself: (1) you're not perfect, and (2) you remain of great value regardless of your imperfections. Having protected yourself from the self-condemnation, spend some time with this hypercritical boss. Engage him in a conversation about his reaction to you. Be open to the possibility that even though you may not appreciate the way that your chief has characterized your weaknesses, he may have clearly identified them. Respond affirmatively to that identification. Draw his specific concerns about you out into the open light where both of you can examine them clearly. By responding affirmatively to this critical but attentive chief, you can strengthen your character and scientific abilities. This permits a renewed relationship with him that is based more on positive responses to helpful suggestions than on an association that is shackled by the links of disapproval and self-condemnation.

Personal Abuse

A chief's unremitting anger and hostility (as opposed to the constructive criticism of the previous example) poured onto her junior colleagues is dangerous and destructive. While it is a truism that you should not have to accept personal abuse in the workplace, it is also true that senior investigators are still provided a wide latitude and are commonly given the benefit of the doubt in these matters. Your institution should have safeguards and procedures that provide at least a modicum of protection for you. Frequently, however, these safeguards are imperfect.

Operating under a cruel chief in a personally and emotionally abusive atmosphere is damaging to you. Working in an environment that is full of anger is like taking a bath in caustic acid; without a good thick protective suit, you'll be burned. Dealing with these painful circumstances requires firm and steady emotional control, qualities that you may not believe that you have, but that you must now develop.

Before we discuss coping strategies that you as an investigator can try, it is only fair to acknowledge that in the end, all the techniques that we will discuss may fail you. If after your repeated efforts you're not able to broker a ceasefire with your chief, you should leave her.

Do not allow yourself to stay in a situation that is personally unhealthy for both you and your character development merely because working with her may be "good for your career."

The belief that you should remain in an emotionally toxic environment in order to keep the best productivity trajectory is self-destructive. There is no guarantee that you'll gain long-term success by subjecting yourself to continued, daily, and unjustified abuse. Since your career is not only your productivity, but also encompasses your principles, your standards, your character, your judgment, your conduct, your ethic, and your temperament, you must be equally vigilant in protecting them all. Staying in a personally destructive environment damages your perspective, alters your conduct, contaminates your professional judgment, and unbalances you. Therefore, if your best efforts fail to resolve the issue, leave.

If you're uncertain of the above guidance, ask yourself what advice you would give to your most special loved one if they were in your place. If your daughter, or husband, or nephew was the object of consistent personal abuse, would you not tell them to leave? Take the advice that you would give this special one because you give it in the spirit of desiring nothing but the best for that good person; that is how you should advise and treat yourself.

Coping from the Inside Out

There are several strategies that you can follow to attempt to work effectively with a chief who is consumed with anger. Our goal, as stated earlier in the chapter, remains unchanged. You as junior investigator should handle this circumstance so that regardless of the outcome (be it good or bad), both your character and your ethic are strengthened.

The foundation of each of the following strategies requires a solid sense of self-worth. Essentially, being the subject of ceaseless anger is the same as coming under attack. However, you meet this attack by fighting it off internally, in your own heart and mind first, before you directly deal with your vitriolic chief.

You must win this internal fight first; victory there ensures that you'll have all of your faculties available and at your command to deal with the abuser. This internal fight can be won by guarding and building your sense of self-value, ensuring that it remains protected and strong.

Specifically, you must reidentify and strengthen your high self-value as the intelligent, capable, principled scientist that you are. Consider your ego structure to be like a foundation for your aptitude, attitudes, and abilities. A good house's solid foundation does not require unrelenting examination in quiet times; however, the occurrence of small quakes and tremors requires that it be consistently monitored, inspected, and repaired. In these times of personal attack, pay critically close attention to your sense of self-value.

> **Your value as a scientist is greater than the valueless abuse being hurled at you.**

Recognizing and reflecting on this truth feeds and adds depth to your character. A firm sense of your own worth will provide the ballast and steadying support that you need for the coming struggles and confrontations. This solid sense may develop in a few days, or it may take several weeks. However, its growth is central to your ability to deal effectively with this growing crisis posed by your chief, a crisis that requires your special care and attention. Take the time now to get strong "from the inside out."

Being strengthened and armed with the recognition that the abuse offers no real long-term threat to you, you can now more calmly evaluate the dynamic of the anger. Is it possible that the anger is situational? Has the angry senior investigator herself experienced a personal shock? The trauma of divorce, the death of a spouse, or new knowledge of the presence of a lethal disease can be expected to perturb anyone's sense of well-being. So can the denial of a promotion, the receipt of heavy criticism, and the loss of institutional support. These are, of course, not excuses that justify the anger. However, if any of these situations are present, they would help to explain the outbursts and beg the question of whether a finite amount of patience on your part is all that is required.

If the anger is not situational, then consider meeting with her to discuss the situation. The meeting should not be an impromptu "hallway" conversation, but scheduled at a time when each of you can discuss this important issue. The purpose of the meeting is to educate her on what she has done and to make it clear that while you're open to fair criticism, you'll not acquiesce to verbal abuse.

If the senior investigator cannot find time to meet with you, then she must be persuaded to make time. We will discuss the advisability of a mentor. Your current circumstance is precisely that situation when this committee can exert his or her influence. Meet with your mentor, clearly explaining in detail your concerns about your boss. Your mentor may have some insight into your chief's behavior based on their experience with her that you do not. After explaining your failed attempts to meet with her, ask that he or she work to persuade your chief to meet with you. If your boss is unwilling to listen to you, she may be open to hearing concerns voiced by her peers. It may be that the meeting must include your boss, your mentor, and you. Be sure to understand any university rules that govern this arena of activity.

Regardless of the presence of your mentors, preparing for this meeting with your abusive chief will require special strength, as she may take the opportunity to engage in further slander.

You'll need your best reasoning and interpersonal skills. However, you're separated from them if you're lost within the tight grip of insecurity and self-condemnation. If this is your state, then your adversary is not your chief; it is your own belief that you somehow deserve unremitting abuse because of who you are. If this is your feeling, then this is a feeling that you must rub out.

Answer the lie that you deserve unearned criticism and punishment with the truth that your natural value and worth is a greater reality than the stressful situation in which you find yourself. Accepting with approval this truth about yourself is central to your ability to communicate effectively in the upcoming meeting.

When the meeting occurs, stand your ground. By this time, you have given important and critical thought to your relationship with your boss. The purpose of the meeting is not to fight with your truculent chief. You simply want to educate her about what she is doing to you and perhaps others around her. You have to find a way to transmit this message in a nonthreatening way. Since every investigator is different, the words and phrases that you need to speak will be unique. However, it is most important to communicate clearly in a nonthreatening way so that she is not antagonized by the phrases you use. In this situation, remember that you're in control if you control yourself.

You might consider saying the following in order to get started:

> Thank you for agreeing to see me. I enjoy working with you and being part of your scientific team. I have some concerns about our professional relationship. If I have some specific deficiencies, I would like to learn what they are so that I can improve them. I am interested in being the best scientist that I am capable of being, and I believe that you can help me do that. However, lately, our discussions do not permit this, and I want to hear from you why our conversations have not been as professional as they should be.

This is submission through strength. Asking in this way demonstrates both your willingness to hear her side of the problem but also serves notice that you expect a professional relationship with her and are willing to work to attain that. Remember, you're not at the meeting to defend your actions, but to educate her about her own conduct. Be willing to accept criticism from her if through this criticism, she is also being educated.

An honest chief who has lost her customary good perspective on her relationship with her junior colleagues will hear the important message that you're conveying. Even though she may not say the words in response that you wish to hear, watch her actions over the next few work sessions to see if you have had an impact on her. Specifically, during this sensitive time, look for signs that she was educated by your conversation with her.

If she has been educated, the relationship may be rectifiable, and you may choose to stay with the team. In fact, recognizing the change in your chief your conversation has generated, your peers may want you to stay on the team as a stabilizing influence.

Sometimes, junior investigators don't react well to this use of their time, thinking that they should be focused on the science and productivity and not on these emotion-laden activities. If you think this way, then I would encourage you to rethink your approach. Your value in collaborative efforts is not merely scientific; it is also to contribute to the collegiality of relationships that is the healthy conduit over which real scientific interchange takes place.

If your conversation was unsuccessful, your chief was not educated, and the character of her relationship with you does not become more professional, then leave her group. Explain clearly to your boss what the reasons are for your departure. Leaving may be uncomfortable for you. In fact, you may have a boss who is vindictive enough to damage your reputation by misrepresenting the circumstances of your dispute with her.

Leave anyway.

20

Investigator as Leader

Seeing Yourself as a Leader

Leadership is the process by which an individual assumes responsibility for a goal-directed project or activity and, through his or her authority, directs the efforts of others to that goal. Some leadership qualities exist in each of us, yet there are only a few people who can (or even want) to develop the right combination of these complex strengths and traits.

At this early stage of your career, it's understandable that you may not seriously ponder the role of leadership as an option for you. The daily whirl of your activities easily requires most of your attention and efforts, with little time and perhaps little inclination for reflective thought on issues of leadership.

The character sabbatical with reflection on whether you could or should be a leader is a fine collection of opportunities to consider and to prepare for leadership.

You may have difficulty seeing yourself as a commissioner of a scientific body, or the head of a research institute, or the president of a university. While this perspective is understandable, consider that many good leaders did not actively seek out a leadership role for themselves.

While leadership roles are sometimes created, they are commonly inherited. Retirements, resignations, and scandal can produce vacuums of authority that must be filled.

On the other hand, you may be intimidated by the prospect, especially when you compare yourself to important historical or contemporary scientific leaders. If this is your concern, it may pay to borrow a page from a politician's book.

When modern politicians consider running for president of the United States, they don't ask themselves whether they are of the same stature as George Washington or Abraham Lincoln. Their perspectives are much narrower. They simply look at the men and women who have or are running for the job and say, "I can do at least as good a job as they can."

On the other hand, one need not be ambitious to develop into a good leader. If you enjoy your scientific career, are intelligent, can be charitable, and are well grounded, you'll have the opportunity to become a leader. With these qualities, you do not have to point yourself in the direction of leadership. Leadership is headed for you.

Traits of Scientific Leadership

Leadership begins with responsibility. To be responsible means that you're liable for the project; you'll be acknowledged for its success and will have to account for its failure. You have something of value to lose if the project fails, and it is this link that ties you closely to the project's outcome.

This tether of responsibility requires that you focus your active attention on the project, extending yourself to work for the project's success. The single recognition that you'll be blamed for, and perhaps hurt by, the project's collapse requires special strength of character, a strength whose growth you can encourage within yourself now as a junior scientist.*

Leadership also requires authority. Leaders must have access to resources and the ability to control the application of those resources (money, equipment, and people) for the project's development and success.

You must accept both responsibility and authority in order to become an effective leader. Responsibility without authority is a recipe that generates frustrating and ineffective activity, frequently producing the project's failure.

---

* Developing this strength is the topic of the next section.

On the other hand, authority without responsibility can produce careless and reckless actions on the part of the leader who has no vested interest in the success or failure of the program.

In order to function effectively as a leader, both your understanding of the importance of the project, in concert with your sense of responsibility, must guide your decisions to commit resources to the project. You must simultaneously consider the value of your resources and their impact on the project's progress.

Understanding your project and your assets then skillfully applying these assets at the right time and in the right concentration for the successful completion of the project requires the delicate touch of a sensitive director.

Meeting the Challenge

The granting of authority is only the beginning of leadership. How you exercise that authority is the true measure of your capabilities as a leader. This capacity is frequently revealed in your ability to (a) be knowledgeable and resourceful; (b) have administrative skill and diligence; (c) have a good strategic sense; (d) be imaginative; (e) possess character, endurance, courage, and self-control; and (f) have a spirit of benevolence.

As you recall, these are the characteristics that we discussed in chapter 2 when you were taking stock of yourself. These traits, along with the ability to be persuasive, are important characteristics of good leaders. However, before you evaluate your abilities in this matter, we must discuss the one characteristic that is the foundation on which leadership skills are based—a solid sense of self.

The Twin Challenges to Leadership

> *No man is so brave that he is not surprised by the unexpected.*
> —Julius Caesar

The fear of failure is one of the greatest enemies of effective leadership, dissolving the will of a director to be successful. This fear, commonly produced from severe criticism or an unanticipated setback, can produce paralysis and ultimately the project's failure. Like acid, it dissolves the bond between the fearful leader and the best parts of his own good nature. Decoupled from both his resourcefulness and his strategic sense, the fearful leader flounders. His capabilities remain, but they have become disconnected and unsynchronized, producing dysfunction and emotional upheaval.

One of the clearest examples of this is the fate of Joseph Hooker.

Appointed by Abraham Lincoln to replace Gen. Ambrose Burnside (who replaced Gen. George McClellan, who replaced Gen. John Pope, who replaced McClellan, who replaced Gen. McDowell and Gen. Winfield Scott), General Hooker was given supreme command of the Union Army of the Potomac in the winter of 1862.

A tall man with striking features, commonly seen astride a magnificent white horse, he was imbued with both confidence and a bellicose nature. His soldiers called him "Fighting Joe Hooker," and he was known to say, "Celerity, audacity, and resolution are everything."

In the spring of 1863, he marched his grand Union Army south into Virginia, hoping to trap and overwhelm the smaller army of Northern Virginia, commanded by Gen. Robert E. Lee.

The movement of Hooker's army was inspired. It maneuvered as one perfectly integrated force, executing complex river crossings with efficiency, moving forward first in one direction and then another. "My plans are perfect," Hooker said, and his men, filled with Hooker's confidence, believed once again in their own abilities to finally bring the war to an end with one smashing victory over an army half its size.

However, when General Hooker, now in Virginia, learned that the enemy was nearby and aroused to his challenge, he abruptly ceased being the aggressor. For several critical hours, Hooker issued a confusing series of orders, converting his aggressively advancing army to a confused collection of units getting in each other's way as they formed an awkward defensive pocket.

When Lee recognized that the invading army was no longer on the offensive, he himself became the aggressor. As Lee divided his army into two units, Hooker did nothing. As Lee moved one of these units behind Hooker, Hooker did nothing. Then at a time of Lee's own choosing, his two detachments fell upon the Union Army. Hooker collapsed, and his army came apart. General Hooker, himself dazed by a cannonball, survived the debacle at Chancellorsville, the greatest Union defeat of the entire war.

When asked later about his curious change in behavior, Hooker responded, "I just lost confidence in Hooker."

The doubt produced by uncertainty, in combination with the burden of responsibility, are the twin scissor blades that severe the confidence of many men and women. However, your recognition of these future threats now, as a junior scientist, wins for you the opportunity to use this early portion of your career to prepare yourself for these twin challenges.

The Genius of Leadership

Anyone who is or has been responsible for a task or activity understands the burden of accountability. The idea of being held to account (and sometimes having to pay a price) for a professional failure is understandably painful to many.

Furthermore, responsibility commonly means not just being accountable for your own actions, but also for the activities and decisions of others. Thus, your career trajectory can be damaged by not just your own actions, but by the misguided and mistaken activities of those for whom you're responsible.

At its core, however, assuming responsibility also means that you're susceptible to a new danger. When the project is not going well, you're the one to blame. Being the point person, you're now vulnerable and easily criticized despite your best efforts. How you're viewed by others is now in the hands of those whose actions you direct since their mistakes will be seen as your mistakes. You're open to attack, an attack that can leave you hurt and damaged.

When viewed in this harsh light, it is no wonder that the idea of responsibility is hateful to so many people.

In many people, much of the pain that is generated by this blame is due to their sense that they have lost value. A leader who experiences a serious setback may believe (perhaps correctly) that others, learning of her failure, will believe that he is not as worthy as he once was.

Perhaps when he quietly comes to grips with her own missteps, she also unwittingly chooses to diminish her value in the face of her defeat, a point of view that amplifies this anguish.

This corrosive process can be unbearable and, ultimately, self-destructive. Fear of this sequence of events can destabilize a leader whose tenure is undergoing a particular period of difficulty.

In addition, this fear of failure is threatening enough to repel those scientists who might otherwise avail themselves of the opportunity for leadership.

The genius of leadership is found in the ability to defeat the fear of failure. The heart of this ability is character development, strengthening your self-worth, and conviction. While failure always remains a possibility, the good leader knows that she will suffer no lasting damage in the face of this setback. Thus, she is able to replace this fear with the knowledge that she will sustain no great injury to herself and her value in the face of defeat. This energizing knowledge permits her to apply her best talents freely and diligently to the project at hand with no core dread of the consequences of blame.

This is the key liberating step that many good leaders are able to take successfully. Having freed themselves from this fear and stabilizing their self-worth, they stay firmly connected to their core talents, drawing upon their best judgment, their best wisdom, and their best vision to prosecute the project for which they hold final responsibility.

This leadership genius does not contradict our previous statements about the necessity of a leader to accept responsibility and blame for a project. Instead, leadership genius recognizes that accepting blame does not mean that you're open to destructive self-condemnation. You accept the blame while ensuring that this acceptance will do you no harm. You'll learn from the defeat, but not have your self-respect and sense of value damaged or diminished by it.

While it is possible to inherit this leadership genius, most of us must learn it. As a junior investigator, the most direct way for you to develop this ability to remove the sting of blame is to decouple your sense of value and purpose from the performance of yourself or your team. This is a direct result of character sabbatical.

Invest in self-significance, not performance significance.

Performance significance has two components. The first is allowing how you feel about yourself to determine your sense of self-value. You either "feel good" about yourself or not. You either "feel like liking yourself" or you don't.

The second component of performance significance is that you permit your state of progress on your work activities to set your self-value. Because each of these components is based on a faulty assumption, your sense of self-worth is impaired, your ability to manage defeat is easily overcome, and in the face of defeat, you feel devalued, beginning a self-destructive spiral.

The snare of performance significance is particularly tight, and overcoming it in the face of defeat requires a monumental effort. As an example of the effort that this takes, consider the following illustration:

> By 1864, Ulysses S. Grant was the rising star for the northern cause in the US Civil War. As a new general, he won clear victories for the Union in the western theater when the poor performance of the northern armies in the east exasperated President Lincoln. During a period when the North lost one major battle after another in Virginia, Grant won decisively at Fort Donnellson, Shiloh, and Vicksburg. These western victories of Grant bolstered his own confidence and sense of significance. In the face of his fine performance, Grant was promoted to lieutenant general, the first officer to reach that rank since George Washington, and given command of all the northern forces. It was expected that with this new title and authority, Grant would repeat his fine western performance in the east.
>
> Being buoyed by this acclaim, Grant rapidly went to work to build on his record of success. He ignored the murmuring of some of his eastern subordinates, who, having observed a string of union generals have their armies and spirits broken against the rock of General Lee's army of Northern Virginia, suspected the same fate was awaiting Grant. Anxious to put these doubts about his performance to rest, Grant launched a new major offensive against Lee's army.

The resulting battle, known as the Battle of the Wilderness, lasted for three days. After intense combat, Grant was handed a spectacular and shocking defeat. The Union casualties were horrific, and the northern newspapers railed against this new abominable leadership of its army. Apparently, the quietly voiced fears were well founded; Grant, like the other preceding Union generals, had been ground up by the sharp teeth of the Confederate Army.

On the third night of the fight, when it was clear to all that he had suffered a major setback, General Grant retired to his own tent for the evening. When the enormity of his failure struck him and he recognized that he was being held up to public ridicule, his taciturn spirit gave way to sobbing.

Hearing this uncharacteristic noise of intense emotional upheaval emanating from their leader's tent, his aides gathered around outside, perplexed by these sounds of intense grief. Some officers, who by this time in the war had seen many men cry, remarked that they had never heard a man break down so utterly and completely as Grant was doing now. They prepared to give the orders for the army to retreat back to Washington, DC, as they had done so many times in the past.

The next morning, Grant emerged from his tent and quietly gave an order. The army would not retreat north but would instead move south toward the Confederate capital of Richmond. For the first time in the war, a staggering northern defeat would be followed by advance, not retreat. The prior night's agony was required for Grant to distance himself from the performance metric and the criticism of others. To Grant, it no longer mattered what people thought or said about him, his performance, or the sacrifices he would call upon his troops to make. Unleashed from a crippling reaction to the criticisms of others, he would follow his own best judgment.

When Lee heard that Grant's army was moving south, he remarked, "I fear that this man will fight us every hour and every day until we are defeated!"

The false assumption underlying performance significance is the belief that your self-worth is not constant and that it can be influenced by your activities. In performance significance, your self-worth is evaluated against the metric of feelings and accomplishments. Since these two metrics are variable, their attachment to your self-worth suggests that your value also fluctuates. Under the rule of performance significance, you have high self-worth on good days and low value on bad days. You're measured by your accomplishments, and since these vary, then so must your value.

However, self-worth is constant. It does not waver with feelings, nor does it increase or decrease with your sense of daily accomplishment. Your worth retains its same constant elevated value regardless of the external circumstances. Your performance can vary from day to day, based on events that are usually out of your control. Feelings are affected by many factors, including but not limited to stress, praise, criticism, hunger, fatigue, hormones, victory, and defeat. However, self-worth remains constant, and your valuation of that worth must also remain immutably high. In a tumultuous sea of anguish and activity, a clear and solid sense of your worth and value that is independent of external circumstances is a solid and dependable anchor for the developing leader.

Therefore, you can defeat the fear of criticism and blame that is generated by leadership with the recognition that their twin impact does not diminish your central value. The reality of your worth is greater than the reality of the criticism and blame that you face as the result of a failure. The acceptance of this reality comes from character development.

Being unafraid of criticism, you're free to accept its constructive aspects and to develop and sharpen your leadership skills because you're not harmed by the disapproval of others. A good sense of your high value allows you to maintain your balance when you sense that your project's progress is threatened. Without a good sense of self-worth, the threat of failure can unnerve you, decouple you from your best skills and insight. Alternatively, the absence of the dread that the project's failure will do core damage to you permits you to respond to a project crisis with alacrity, calling on the best of your vision, sense of timing, and judgment as you apply resources to get your project back on track.

Defeating the Peter Principle

A maxim from one of the more famous (or infamous) books of the 1960s states,

> The Peter Principle: In a hierarchy, every employee tends to rise to his level of incompetence.[1]

This principle was the result of a study that examined possible reasons for the high levels of incompetence observed in many corporations. In prominent companies, incompetent employees were present throughout the corporate structure regardless of rank or the degree of expertise required for the job. The explanation offered for this presence was that the prevailing corporate culture promoted employees until they were "in over their heads." These promote demonstrating competence at the lower rungs of the corporate ladder, repeatedly earned promotions until they reached a level at which they could no longer demonstrate the proficiency needed for the task at hand. The demonstration of this new incompetence ended their continued rise through the corporate structure. However, it did not lead to their termination or demotion. They were permitted to remain at this level in spite of (in fact, because of) their incompetence.

Belief in the Peter Principle can underlie misgivings about a new leadership opportunity presented to you, especially if you're doing well in your current job. Why should you be willing to put yourself at risk? You, like all junior scientists, do not wish to fail, just as hardworking corporate employees don't wish to be incompetent. However, when promotions are offered to employees, the employees commonly accept them in good faith. The employee is being rewarded for the clear demonstration of their competent efforts, and the reward is offered by a management group that believes the employee will succeed in the new higher position. Ultimately, the promoted chooses to believe that she will succeed as she has succeeded before, that is, through the sustained application of her good practices, skills, and work habits.

It is this last comment that undermines the ability of the employee to successfully rise and may interfere with your rise to a leadership position. Promotion requires not that you practice your old skills at a higher level, but that you instead learn new skills. Competence in your current job means that while you have demonstrated the ability to perform well, even admirably, you also may have become tolerant of your weaknesses. These weaknesses continually exert their influence, but over the months and years, you have learned to work around them. You have made room for them and be comfortable with them.

> **Promotion requires not that you practice your established skills at a higher level, but that you instead gain new ones.**

These weaknesses, while not doing any tangible harm in your current work activities, can cripple your efforts in the new job. The Peter principle assumes that you're limited in your ability to overcome your weaknesses when you accept a new demanding position.

Therefore, when you take on a new position of leadership, you must be willing to critically review your weaknesses, with the view to converting them to strengths. Relying on your natural skills is not sufficient to meet the new challenge; you must develop new ones.* You don't have to know how to do this. You must simply be willing to do it.

Begin by assessing yourself. While appraising and undoing your weaknesses, take some time to evaluate your strengths. The talents that you had in your former job are in all likelihood not enough for your new tasks, and you'll need to develop new skills. This can be a delicate matter. You have earned your new position by demonstrating (and essentially relying on) your capable skills. To advance to a new position successfully, you'll need to look at those skills critically with the goal of retooling them and developing new ones. In a sense, you must deemphasize the talents that have permitted you to rise to this level so that you can develop new ones.

You grow into leadership by developing new strengths and leaving old weaknesses behind. Expose yourself to new concepts and ideas.

The secret to defeating the Peter principle is self-transformation and growth. Being free from the fear of being damaged by failure, you can extend yourself into new areas, working with new energy and diligent effort to develop new strengths that are not "natural" for you. Specifically, this means that you should (1) acknowledge that you have within you the strength to learn the new skill (even though you may not know how) and (2) utilize that strength by applying the best of your knowledge, training, experience, and expertise to the exercise. If you have

---

* Discussed in chapter 1.

always been quiet and shy, take the opportunity to extend yourself into first limited and then expanded regality. If discipline has been a problem for you, take the time to gain self-control. If you're loquacious, then devote yourself to governing your tongue.

The key to defeating the Peter principle's hold on you is your directed energy to shatter your own self-limitations in the face of promotion.

Leadership Traits

With your source of self-worth secure and the recognition that you need to learn new skills acknowledged, we can now discuss some of the traits and characteristics of good leaders. Execution of these talents in the right combinations is what allows the team to make good progress. Experience provides the intuition and the sensitive feel of action in applying leadership skills in science.

The traits that are discussed in this section are not meant to be exhaustive, but merely demonstrative of the tangible abilities that the outside observer identifies in good leaders. As you work to develop these traits, try to avoid the temptation of focusing on one at a time, that is, "Today I will try to be charitable; tomorrow will be my 'resourcefulness' day." Instead, let the blend of these principles accrue and develop naturally.

Develop an attitude that prepares you to demonstrate each of these leadership talents during your activities. As this attitude becomes easier for you, you'll be able to wield these traits in very effective combinations.*

Knowledge and Resourcefulness

Resourcefulness is the ability to use the material and personnel that you have in the most effective way to solve the problem that faces you. Resourcefulness requires that you do three things. First, know your problem. Second, know your resources. Third, be driven to solve your problem.

Knowing your problem specifically means that you must completely understand the dilemma. What is its genesis? Why is it an obstacle that blocks your way? Have others faced your problem? What solutions have they tried? Why did those attempts fail? If phone calls to authorities would be helpful, then make these important contacts. Do anything that you can to learn about your problem. Become the problem's expert.

---

* You'll see that several of these characteristics were covered in earlier chapters. In these situations, reference will be made to the earlier discussion.

Secondly, know all about your resources. Your resources are provided for you to solve the problem. However, they cannot be used effectively if you don't understand their strengths and their limitations. Become acquainted with them all. If you have computing resources, then understand what these machines are and are not capable of. Completely familiarize yourself with the instrumentation. Know what they do, how reliable they are, and how much they cost to use.

As essential as it is for you to learn about the material resources, it is even more critical to understand the personnel who are under your direction. Learn their strengths and weaknesses. Look beyond their job descriptions. Who are these people? Who among them is the writer? Who is the debater? Who has the technical skills? Give them relatively small and simple problems to address, allowing you to gain an important sense of their ability to interact. In addition, take the additional step of helping them to overcome their weaknesses. Make the commitment not just to the project's success, but to your staff's development as well.[*] You may find that you do not have enough resources. If that is the case, do not hesitate to ask for more, with the full intent of applying these new resources to the problems at hand.

Finally, you must want to solve the problem. The desire to solve the problem is the source of the energy that will drive your repeated efforts to unravel the issue. Attempting to solve the problem means assembling and reassembling your resources in one intelligent combination after another as you try and retry to resolve the issue.

Like finding the right combination to the lock of a treasure chest, your desire is the energy that fuels your search for the right solution. If you don't really want the prize, you'll give up after a few half-hearted attempts.

Imagination

Difficult problems are commonly perceived as such because they defy the common approaches that work so well in the solution of simpler problems. The solution to a tough problem commonly requires ingenious and innovative ideas.[**] It requires imagination.

Commonly, imagination is perceived as a gift that is, in some mysterious fashion, genetically or otherwise conferred. Many people

---

[*] This was discussed in detail in chapter 3.

[**] This is colloquially expressed as "thinking outside of the box."

believe that "you either have imagination or you don't." This is generally not the case. Imaginative solutions are the result of the combination of information, intuition, and the drive to solve a problem. Knowledge leads to intuition, and intuition plus desire produces the insightful, creative solutions that commonly are considered to be the product of imagination.

Specifically, imagination is developed from repeated attempts to apply novel approaches to the problem at hand. In order to be productively imaginative, you must have solid knowledge of the problem, as discussed in the previous section. However, you must implement this knowledge in new ways. One such example of an approach might be to ask if the problem can be broken into smaller pieces. If these smaller pieces can each be solved, then perhaps the final solution can be created from the assembly.

Another useful maneuver in solving the problem is to attempt to discover an assumption that, if it were true, would make the problem easy to solve. Once you have identified the assumption, apply your resources to make the desired assumption a reality. Having accomplished this, the problem will be more easily solved. First, solve the problem any way that you can then solve it the best way that you can.

There is a mental flexibility and alacrity involved in trying these approaches. The primary energy source for this cerebral activity is your drive and commitment to identify a solution.

Finally, the ability to think imaginatively is an important dividend of having a secure and separate source of self-worth. If you fear the blame of failure, you're not likely to take the risk of proposing an untested and imaginative solution. Cutting yourself loose from the fear of failure frees you to develop ingenious and innovative approaches to your team's complicated problem.

Keeping Perspectives Straight

In leadership, you have to successfully manage both the "big picture" while staying in close contact with the daily activities that are essential for the continued operation of the project. This can be complicated because the management of these two issues can require contrary solutions. Specifically, the successful resolution of a short-term problem can lead to the unraveling and destruction of the long-term plan. This sometimes requires important redirections of goals and objectives, itself a complicated maneuver.

There is perhaps no more visible example of the complexity of this issue than observing the tasks of those who manage campaigns for the various candidates of the major parties for the president of the United States. Gaining this high office requires that the candidate first win their party's nomination then use that nomination as a springboard to the presidency. However, since the 1960s, winning the party's nomination means that the candidate must satisfy the needs of those who vote in the primary elections. If the candidate is running for the Democratic Party nomination, he traditionally must satisfy primary voters who tend to be more liberal. If the candidate is running for the Republican nomination, then he typically will have to satisfy voters who are likely to be more conservative. However, once the nomination is gained, each candidate has to appeal to the vast majority of voters who tend to be in the middle. Here is the difficulty. By working too hard to achieve the short-term goal of their party's nomination, the candidate runs the risk of alienating the voters in the center whose support they will need in the general election. Additionally, the candidate's postnomination "move to the center" can alienate the voters who supported the candidate in the primary elections. The management of the short-term (party nomination) versus the long-term (national election) goals is a delicate operation.

While the issues are somewhat less climactic in science, the need for both clear vision and a deft touch in steering the team is essential. Sometimes a step or two away from the long-term goal is required to reach the end. Like a pilot who steers an airplane around a storm to reach the destination safely, the best path may be a circuitous one. In this matter, there is no substitute for a good view of the goal.

Administrative Skill and Diligence

Administrative skill teaches you how to avoid wasting time. This might be considered an unusual statement because many junior investigators believe that their involvement in administration *is* wasting time. However, the effective assembly of your resources to solve your problem requires that you understand where they are, what state they are in, and what their needs are. You may have developed a good action plan of action to solve a scientific problem involving the creation of a database, only to learn that the programmer whose contribution to the project was key is scheduled to take vacation during a critical phase of the production. Simple but diligent attention to this type of issue early in the project's development would avoid this critical and unnecessary delay in your project's completion.

There is no getting away from administration. You either spend a little time up front dealing with it or spend critical time later in the project trying to catch up. Prompt and sedulous administrative attention can save the project a good deal of wasted time down the road.[*]

These next two traits separate leaders from "managers."

Boldness and Achievability

Boldness is the force that activates and energizes the imaginative idea. Rather than a wild hunch, boldness is based on shrewd reasoning and a calculated risk.

One of the differences between a good leader and a good manager is that the leader has a touch of the gambler in her. She is willing to take a risk that can produce a major advance for her team. However, taking a risk is not the same as engaging in a reckless enterprise. A good leader puts her leadership and her team at risk only after a complete exploration of all the ramifications. The most productive bold activity is based on the most detailed plans and considerations. The resultant actions are not based on immature snap decisions but are instead rooted in solid motivation and well-considered plans.

> In the winter of 1776, the American Revolution had effectively ended as a victory for the British Empire. In the six months that followed the Declaration of Independence, Washington's army of eighteen thousand had been whittled down to five thousand men. No longer an effective fighting force, greatly outnumbered by a well-organized and competent British/Hessian army commanded by the confident General Howe, the American force managed to escape south to the Pennsylvania side of the Delaware River. Rather than pursue this ragtag group at the beginning of a cold winter, Howe let them go, believing that the bitter and frigid northeastern weather would finish off the rest of the American "militia."

---

[*] We have had much to say about administrative diligence in chapters 3 and 4.

However, Washington, after retreating from the British for half a year, also realized that encamping his army for six months in despicably frozen conditions would shred it. Having a touch of the risk-taker in him, he decided to do the unexpected and attack. Recognizing that there were nine hundred Hessian soldiers encamped in Trenton for Christmas, Washington took half of his five thousand men and recrossed the Delaware to attack these German soldiers on Christmas Day.

The American assault was completely unexpected and totally successful. Not only was the battle won with minimal casualties (two Americans froze to death on the ride across the freezing Delaware River), but his victory electrified the American cause. The countryside, heretofore wary of the revolutionaries, opened their homes and supplies to them; and volunteers flocked in to join Washington's army.* The British, on the other hand, were more than embarrassed. They realized that the war would not end quickly and would require the one thing that neither the Crown nor its subjects had in this matter—patience.

Washington's decision to attack an unsuspecting force was not a rash action, but a well-considered and carefully executed gamble. Bold ideas have to be planned with great care and meticulous attention to detail in order to accurately gauge the likelihood of their success. Washington knew his adversary and their Christmas Day habits. He understood the limits to which he could push his men. He successfully discerned whether the denizens of the surrounding countryside could be trusted not to warn the Hessians of the predawn American advance. While Washington's imaginative idea was sparked by boldness, it was guided by the very best discipline, planning, and judgment.

Without audacity, coupled with a sense of the achievable, a leader is simply a manager. While it is certainly appropriate for the young scientific leader to fear rashness, she must also avoid irresolution.

---

* Many of the volunteers from Pennsylvania joined Washington's army because they were delighted at the prospect of invading their New Jersey neighbors!

Take Advantage of Your Mastery

New science is fueled by new concepts and new technology. Each of these can offer important intellectual hurdles. However, each contains critical material that might be applied effectively by your team. In order to assess the potential of these important contributors, you must understand the new ideas and/or the new technological advances. Although this is difficult material, a good leader will insist on learning it.

More than this, a good team leader will not just learn the material, but he will also become skilled in the subject. The new technology offers a steep learning curve. The leader will acknowledge it and successfully climb it. He will understand the topic in all of its subtleties, complications, and implications. He will comprehend them so that he can explain its salient features to his team, explicating it in such a way that his team members are able to absorb all that they need to in order to apply the material to their role on the team as well. This is not just learning the material—it is mastering it.

However, learning the information for the sake of learning it does little to advance your team's progress. Once the science is mastered, move rapidly to make new progress on your project based on what you have learned. Seize the opportunity that your mastery of the new material has earned.

What commonly happens is that although a group of leaders is able to understand the utilization of a new technology, they are unwilling to extend themselves to take the next step of application. They absorb the new material—then wait. Acting as though someone or some organization will seek them out after the leader and his group have learned the new material, they do not aggressively seek out the new opportunities that this knowledge has gained for them.

While you can take comfort in your mastery, do not languish in it—take advantage of it. If after mastering the material, your scientific team, through their learning and their own disciplined review of that learning, has developed a new concept, then move rapidly to get this concept out into the mainstream of your field. Move aggressively to identify a venue that would give you the opportunity to discuss your idea. Look for and seize the opportunity to write manuscripts that allow you to disseminate your idea. While learning is important, it is not the only important thing. Developing ideas based on what you have learned is also important. Don't wait for the sake of waiting. Consider the following illustration:[*]

---

[*] Ackerman T. and Berger E. "Deal brings rivals to the same table. Hospitals

In 1951, Dr. Michael DebaKey, preeminent cardiovascular surgeon in Houston, Texas, offered Dr. Denton Cooley a position at Baylor College of Medicine. This offer permitted Dr. Cooley, who had just completed his training in Baltimore, to return to his Houston home base. Dr. Cooley's arrival ushered in a period of productive collaboration for these scientists. Year after year, the two surgeons worked together as they taught, conducted research, and developed new devices. However, after ten years, Dr. Cooley left to form the Texas Heart Institute. While remaining close, the two famous surgeons chose to no longer operate together.

In 1965, DeBakey participated in a federally funded project to develop an artificial heart. After many months of work, the project was near completion, and several physicians advocated that this first artificial heart was ready to be studied in people under the most tightly controlled circumstances. This next step would have been a first in human surgery, and the successful development of this device would have, in the eyes of some, been the event that would have earned Dr. DeBakey a candidacy for the Nobel Prize. However, DeBakey demurred, continuing to work on his device.

To international acclaim, the first artificial heart transplant took place in 1969. With great care, the artificial heart was implanted in a forty-seven-year-old man who was awaiting a heart transplant. The patient survived for sixty-five hours, dying after the artificial heart was itself replaced by an organic heart from a donor. The research community reacted with thunderous applause at this tremendous surgical feat. However, the cardiovascular surgeon who was responsible for, and who carried out the surgery, was not Dr. DeBakey, but Dr. Cooley.

Dr. DeBakey accused Dr. Cooley of misconduct. Stating that the artificial heart Dr. Cooley had transplanted into the patient was identical to the one

partnering but Cooley, DeBakey might not." *Houston Chronicle.* Friday April 23, 2004, page 1.

under development by Dr. DeBakey, Dr. DeBakey stated that Dr. Cooley had implanted the device without Dr. DeBakey's permission. The American College of Surgeons chose to censure Dr. Cooley, and after an argument with the trustees of the College of Medicine, Dr. Cooley resigned from the institution.

The inability to retain the initiative by one, and the unethical conduct by a second, further corroded a relationship that had been so productive for cardiovascular research.

While part of the reason for being quick to take advantage of your team's mastery of a topic is competitiveness, there is a deeper motivation. Moving forward with the idea requires good clearheaded thinking about the application, deep thought that might not be so incisive in the absence of a public review of your work and ideas. Secondly, moving these ideas into the open spurs the ideas of others while clearing them from you, allowing you to move on to newer important work.

Erring on the side of caution is an inadequate excuse for the lack of boldness and resolve in a leader.

Developing Other Leadership Traits

It is rumored that Napoleon Bonaparte, at the height of his reign over France in the early nineteenth century, was asked why the world did not have more good leaders. He replied that a good leader was required to have disparate and contradictory traits. During the rise to leadership, the candidate had to be single-minded. He had to be venal, merciless, dishonest, conniving, and malevolent in order to gain power.

However, once the goal of leadership was attained, the requirements changed. The new leader, in order to be a good one, has to replace these unenviable traits with other more enduring characteristics. Among these are were open-mindedness, generosity, inquisitiveness, honesty, and kindness. Since few people could make the transition, Napoleon reasoned, the world was bereft of good leadership.

If Napoleon said this (and he was certainly in a position to know), then perhaps a useful modern-day interpretation of his statement is that people have to grow into good leadership. Leadership requires self-transformation. It requires leaving old weaknesses behind while you develop new strengths. The ability to develop these traits is directly related to your desire to develop them. While you cannot in general just think yourself into a good character, you can develop the depth of desire to improve your character and then let that desire drive you. As someone once said, "It's not the skill, it's the will!"

What specifically does this drive compel you to do? Let it compel you to meet and speak with people who you believe have good character. Let it compel you to read about people of good character. Let it compel you to discuss these people with others who would be willing to listen to you. By these activities, you begin to embed yourself in the sense of standard, sensitivity, and judgment of those men and women of character, opening your own character up to be altered and shaped by the experience. The good character that develops leads to strength. This strength produces self-control, and self-control produces both courage and endurance.

Source of the Best Ideas

As a leader, you have the power to command, a power that has been granted to you by a legitimate authority, be it a department chairperson, a dean, stockholders, or a legislature. However, the art of leadership is not the brusque exercise of this authority.

Abraham Lincoln said that the art of politics was the ability to get people who do not like you, and who want to follow their own personal whims, to do what you want them to do. The closer the option of commanding by fiat is, the more delicate the art of leadership becomes, and the greater the need for discipline. Thus, the beginning of the art of leadership is learning not to use what is available to use. Do not lead by force of authority.

To develop this skill, you might consider how you would lead without any sanctioned authority. If you were a member of the team and not a team leader, how would you convince others to follow your direction? By persuasion. The ability to persuade is one of the most effective traits of a good leader.

As the leader, you begin with the recognition that your team members are themselves intelligent adults who, like you, feel the need to understand the relevant information concerning a decision you must make. In the ensuing discussions, they would like to provide and you would need to listen carefully to their input and observations. However, as their leader, do more than merely listen. Listen with the attitude that your position can be altered and influenced by their observations and opinions. Let your team know that you're open to modifying your own point of view based on their observations and reasoned conclusions.

After your team members have each had the opportunity to provide their own opinions, challenge them with alternative ideas. Let them listen and respond to the criticism that others might bring against their perspective. Help them to think critically about their point of view while keeping yourself open to accepting their modified position. Encourage your team to develop confidence in identifying useful observations, ideas, and suggestions. Listening to this advice in the spirit of acceptance allows you to absorb the best of their ideas for the good of you and the team.

There are two advantages to this approach. The first is that after all, a team member can have a better solution than you do. One of the best ideas that a leader can have is that they, as leader, do not always have the best ideas. Commonly, the preeminent concepts are generated by the team members themselves.

Secondly, and perhaps most importantly, you transmit to your team the attitude that you're willing to adjust your point of view based on what they have to say. This is the central piece of persuasion. By being willing to adjust to their attitude, you can create in them a new willingness to adjust their position toward yours.

After having carefully listened and discussed with your team their own perspectives, provide your own position that has been modified by the conversations that your team has just completed. Point out to them exactly where your own point of view has been altered by those of your team. Show them where their points of view have been incorporated into your own. Combine tact, prestige, firmness, and character to appeal to the better natures of your wavering team members.

After having all points of view in hand and completing your own deliberations, make the required decision, and follow this by explaining your rationale. This clear explanation is important even if (and perhaps especially if) it is not obligated. Your willingness to extend yourself to explain your answer demonstrates your concern for the point of view offered by the team members who may disagree with your decision. Even though you disagree with them, be willing to explain how you came to your decision.

Despite your best effort, you may have to make a decision in the absence of unanimous support among your team. In this circumstance, make the decision that you believe is in the best interest of your group, but make the decision in a way that keeps the team together rather than in a way that creates fissures. Begin by recognizing that those on your team who believe that you have not decided correctly will nevertheless have a very useful perspective and ultimately may be proven right in the end. It is this group that will most likely be among the most vigilant for observing signs that your decision was incorrect as the group moves forward with implementing your decision. By staying in close contact with them and listening carefully to their observations, you can use their perspective as a useful early warning if your plan is not producing the result that you anticipated. Alternatively, if your decision was ultimately correct, by remaining observant, your critics in the group will be among the first to see it.

Finally, be ready to apologize to any of your team members. If it turns out that the individuals whose point of view you chose not to accept were right, be sure to acknowledge that to them. Learn how to apologize from strength and not from weakness. To apologize from strength means that although you were wrong, that in no way diminishes your worth or value since these are independent of your achievements and mistakes. Your apology can be earnest, sincere, simple, and deep, with no motivation other than to be certain that the fact that you were wrong is understood by others. Alternatively, apologizing from weakness occurs when your self-significance is linked to how correct your decisions are. Such apologies can become complicated because the requirement of acknowledging that you were wrong becomes entangled in your need to help to regain your loss. In this case, the apology is converted from the simple sincere admission that you were wrong to an attempt to soothe the pain of self-diminishment that occurs as a consequence of linking your value to your performance.[*]

The Genesis of Team Leadership

While the different components of leadership are discussed earlier, one of the key features of leadership is administrative diligence. Without the ability to successfully administrate a research team, your capacity to be responsive to that team's logistical needs is impaired. As a consequence of this defect, the ability of your team to produce its best science will be blunted because it cannot bring to bear the resources that it needs to create the best scientific product. In this section, we will discuss the administrative side of leadership with specific focus on research team management.

---

[*] Discussed earlier in this chapter.

While you may get the opportunity to work on the science of the research project regularly, you'll have to come to grips with the administration of the grant on a daily basis. This includes, but is not limited to, communications with the agency that funds your grant, discussions with coinvestigators who work in distant research centers, discussions with your research team, conversations with animal and human protection committees, as well as communications with recruitment committees, ancillary study committees, compliance committees, core lab committees, and publication committees. You'll also (directly and indirectly) supervise the work of administrative assistants, secretaries, information system operators, and computer specialists. In order for your project to function smoothly, these specialists must have confidence in their own ability to support the research project.

As in any activity, there will be important questions and conflicts that arise. In these predictably common and potentially tumultuous circumstances, your colleagues and support teams will look to you to resolve those conflicts and to keep the research project moving in the right direction. Whether a computer programmer or a secretary can keep their jobs (and support their families) will depend on your ability to carefully and skillfully manage the financial aspect of the grant.

"When in Command . . . Command"

The prospect of being principal investigator (PI) can be intimidating. Sometimes it arrives unpredictably, for example, through the unanticipated demise or departure of the current PI. However, when the opportunity presents itself, first seize it,* then apply the very best of yourself to this new challenge.

Defeat comes more from the fear of failure than from taking a bold and calculated risk. Consider the following example:

> Chester Nimitz was the commander of the United States Naval Pacific Fleet immediately after the US major defeat at Pearl Harbor in 1941. Following this debacle, Nimitz's major responsibility was to maintain a naval presence in the Pacific Ocean until the United States could recover from its losses of men and equipment. Sensing that the enemy was moving east across the

---

* This assumes that being principal investigator lies along your well-considered career trajectory.

Pacific Ocean toward Hawaii, he chose to face them at Midway Island.

Nimitz's strategy of choosing to defend this little-known island that was far from the expected line of enemy attack rather than placing his remaining forces to protect the Hawaiian Islands ran counter to the opinions of senior leaders, generating severe vocal criticism of him by the governing powers in Washington, DC.

To make matters worse, Nimitz's senior commander, Admiral Halsey, was stricken with a debilitating skin disease, effectively removing Halsey from any role in the upcoming battle. When Nimitz asked Halsey to name his own replacement for the upcoming confrontation with the enemy, Halsey immediately replied, "Ray Spruance."

Nimitz knew Ray Spruance. A promising younger officer, Spruance had demonstrated his skill at handling smaller ships in battle. However, Spruance had never commanded a naval task force; and no one knew whether he possessed the character, temperament, and faculties to effectively maneuver a large flotilla of ships of different types in the heat of battle.

When Nimitz, visibly uncomfortable, wondered aloud about the negative reaction to this controversial appointment from political leaders who were already critical of Nimitz's decision to defend Midway Island, Halsey calmly replied, "Chet, when in command . . . command."

As a junior investigator, you'll most likely have the opportunity to manage a small grant that has a microcosm of these problems. However, the management skills that you master in small research activities will serve you well as you progress to the administration of larger projects.

Guides for Management

We have pointed out the plain truth—administrative support and sustenance are the tracks on which the research train runs. Just as the train will not reach its destination if the tracks are twisted or bent, your research grant cannot proceed unless its administration is smooth and predictable. Specifically, you'll not be able to carry out the science of your project (that was the reason that you applied for an accepted research role) unless you can get its administration running easily and efficiently.

We have also affirmatively stated that the successful execution of both the administration and the science of the project has been placed squarely, unavoidably, and undeniably on your shoulders.

Completely accepting this onerous responsibility is your first major task as principal investigator. Carrying out the related activities is, however, relatively straightforward. The successful administration of the research project requires your time and your knowledge. Specifically, in order to successfully administer the grant, you must do the following:

1. (Take the time to) know the subject matter.
2. (Take the time to) know your people on the grant.
3. (Take the time to) effectively communicate.

If you're willing to devote the best of your time and effort to these activities on a regular and frequent basis, project management will, by and large, be straightforward. However, consistent and patient attention to each of these three tenets, will bring you closer to the needed solutions to complicated administrative problems. We will consider each of these in turn.

Know the Subject Matter
The foundation of the research project is scientific knowledge. Essentially, the administration of the research project is the delivery of the right combination of financial, technological, and human resources to propel the scientific work.

Administrative diligence begins with superior knowledge about the science of the research grant. Whether to hire an additional project coordinator or a second programmer is based on the scientific needs of the project. Whether an additional receptionist is necessary may depend on whether more investigators are needed for the project, which in turn is based on the project's underlying science. The administrative decisions are determined by what benefits the project's pursuit of the science. Therefore, a complete understanding of the underlying science is critical.

The requirement of this knowledge mastery by the principal investigator may not be as self-evident as it first appears. Certainly, you as principal investigator, are expected to understand the scientific content of the research that you're directing. However, understanding the technical details of just one component of the grant, leaving the rest of the details to others, with only a broad and general understanding of the science in these other areas, is not sufficient. You must be conversant with it all.

Discussions about administration are commonly more about the details of the research project than they are about its overarching theme. As the science becomes more technical, more specialists are necessary to conduct the research. An anticipated and natural consequence of this is the distribution of the knowledge base among several scientists. The problem this fragmentation produces is that, when an administrative problem arises, its solution requires different and disparate facts about the project. One person does not know all the facts, but a combination of people do. Thus, the conversations necessary to solve these problems have to be delayed because either (1) the required scientists with the critical information cannot get together due to interminable schedule conflicts or (2) the scientists are unable to successfully integrate the required pieces of information that they have because they do not understand (or are not willing to take the time to understand) the administrative problem. Thus, the administrative problem, like a nagging toothache, becomes a chronic issue, growing into a larger and larger distraction from the research effort. However, if the principal investigator takes the lead in mastering the knowledge base of each component, many of these administrative difficulties can be quickly resolved because she has already absorbed the requisite knowledge.

A solid knowledge background is one of the pivotal quantities on which leadership skills are built. Mastery of this material requires a great effort. A research project that contains a collection of clinical, genetic, and quality of life components that will be entered into a central database requires the principal investigator to learn important details of each of these components. The principal investigator should also understand the data entry component. Learning this material is not hard; however, it does takes time.

For example, if you have no computing or programming background, but you are the principal investigator of a project that requires these skills for either database development, data entry, or data analysis, then you should plan to spend several hours over a relatively short period of time discussing database architecture with the information technologist of the project. Work with her, asking her the same questions repeatedly and in different ways until you understand what it is that she is telling you about the project's development.

There are two benefits that come from this effort. First (and obviously), you gain a new understanding of the use of a database for your project. When related issues arise in discussions of the project, you'll have a new knowledge base from which to draw, as well as the confidence that comes from gaining new critical information about an essential component of your research project.

In addition, you're viewed with new confidence by your research team as you demonstrate your commitment to the project since this knowledge directly translates into a refined sense of the goal and direction of the project. Finally, you gain the appreciation of the information technologists since you have demonstrated the importance of their commitment to this project by affirmatively choosing to spend time with them, allowing them to explain their work to you. This commitment can pay handsome rewards later in the project when they may be expected to "go the extra mile" for the good of the study.

In addition to learning the details of the many different scientific areas that are essential to the research project's success, it is important to integrate the scientific and administrative components of the project. There are timelines that must be satisfied for the scientific progress of the study. The arrival of specimens to a core lab must be timely in order to complete one component of the research. Data monitoring meetings must be scheduled and planned. Meetings of animal and human subject committees must be folded into the research execution. Data will be required for these meetings. Therefore, data will have to be collected, entered, checked for accuracy, analyzed, understood, and presented.

If each of these activities is not carefully considered and planned, they tend to merge into a blurred mass of events that will continually ambush your research schedule. In addition, there is always the unexpected.

The solution to this aspect of the administration is an investment of your time. There is no substitute for the careful and methodological planning of the numerous research-related events that can be scheduled. This preparation requires the junior researcher to develop and improve his ability to correlate and coordinate these many activities. There is no doubt that this careful preparation is time-consuming. However, since careful forecasting is necessary, devote an important component of your time to it so that your plans are accurate, reasonable, and executable.

If you have a portable computing device, affirmatively use it to help you schedule and track these events. After all, if these devices are good for anything, they are good for keeping track of dates and deadlines. In addition, handheld digital recorders are available to log and save spontaneous verbal notes and memos to yourself. If you get an idea or remember and issue but cannot take the time to write it down, use the recorder, playing it back later.

Improve your management skills by actually mapping the timeline of your research in detail. Lay out each component and meeting that the research effort will require for the foreseeable future of the research scheme. There is software available that allows you to do this on personal computers. There is even software that will run on a PDA. If you're comfortable using electronic devices for these planning activities, this software can be invaluable to your planning.

There are two advantages that you derive from creating this research activity map. The first is that you'll have a legible record of the progress of your research with its many deadlines against that you can more objectively gauge the process of your research. Secondly and more importantly, engaging in this process demonstrates to you the limits of your own knowledge of the research project's activities. By developing the map, the information that you do not have but require in order to complete the schema of activities and deadlines becomes clear to you. New questions arise as you review the timing of meetings and the data that you'll be required to supply at these meetings. Choke points in the research activity will also become apparent to you as you begin to lay the activities of the research out.

When the map is complete, just don't set it aside. Continually appraise and update it as time and your research project activities advance. Take the time to update your task map as new developments occur. Schedule and vigilantly protect the time you need to review the progress of the research in all of its aspects.

By devoting the time to master the subject matter of your research, you place yourself in a perfect position to efficiently, accurately, and benevolently make decisions about resource availability to keep your research project on track. Examined another way, the maximum size of the research project that you can constructively manage is, to a great extent, determined by the degree to which you can master the contents of the research's subject matter.

Know Your People

In all likelihood, although you, as principal investigator, hold the ultimate responsibility for the grant, you're not its only worker. There will be coinvestigators, information specialists (i.e., data entry and or computer programmers), a business administrator, and perhaps a receptionist or secretary. Each of these workers commits some of their time to your research project, and each has skills and abilities that are different from yours. These team members will work hard because of their professional competence even though they may not derive as much credit from the grant as you.

New Hires

While some workers may already be on the job, it is not uncommon for you to find that you need to hire additional people. This can be an intimidating prospect for the junior researcher for two reasons. The first is that the skills that you're looking for are very likely skills that you do not have. For example, you may be required to hire a computer programmer with skills in modern database management. With no such skills yourself, how can you satisfactorily test the knowledge base of the individual that you're hiring?

Secondly, you may not have ever hired anyone before in your life.

Try to avoid being overinfluenced by an impressive CV. Sometimes, important weaknesses in the interviewee's personality or character can hide behind a thick résumé. At a period in my career when I owned a physician's association and had the responsibility of hiring physicians, one physician I considered for the job had the most impressive curriculum vitae that I had ever seen at the time. In addition to training in medicine, this physician was well versed in both the classics and the hard sciences, having attained a PhD in physics. The CV was over one hundred pages long. In fact, it was the only CV that I have ever seen then (or have ever seen) with a table of contents and an index. This CV was not a résumé—it was a publishable autobiography! I hired him, a choice that was based to a large degree on the persuasive power of this impressive document.

Unfortunately, my decision to hire this physician was one of my worst hiring decisions. Within a few short weeks of his employment, it became clear that he was unable to manage patients using the appropriate and accepted standards of the community. Ultimately, he had to be removed from his job for incompetence, an action that produced a hail of angry and litigious correspondence from the discharged physician. Judging that book by its cover led to disaster.

Rely on the expertise of others to judge the competence of the individual. If you're not able to assess the skills of the candidate, allow your colleagues who have such experience to interview the candidate and provide an assessment. Sometimes, you may have to borrow a colleague from another project who is in a better position to assess the abilities of the candidate than you are. This requires that you, at the very least, be able to clearly articulate the needs of your research unit.

Secondly, develop the skill to conduct a balanced and fair interview process. Make sure the candidate understands the requirements of the position that is being offered. Have several people interview the candidate. As a useful metric to help you develop your interviewing skills, find someone more senior than you with whom you can jointly conduct the interview. You can learn a great deal from watching the senior person conduct their component of the evaluation.

During the interview, be sure to give the candidate an opportunity to fully express their goals, their skills, and why they are a good candidate for the job. If you do not have the expertise the job requires, then encourage the interviewee to explain in simple and plain terms what they will do to accomplish the task. If they can explain this clearly and the approach can be validated by others who have experience in the field, you may have discovered a fine technical person who also is communicative.

Finally, be sure to be patient. A diligent, capable, and fair-minded search commonly does not yield quick results, but does commonly yield good and lasting ones.

Access

Alternatively, the cadre of workers that are already part of the project requires your attention as well. The clearest demonstration of your support of them is your time commitment to them. As professionals, they will work hard without it, but administering this grant becomes much easier if they have regular access to you. An atmosphere where everyone contributes for the good of the project doesn't derive de nova but must be created. It is most simply and directly generated if you as the principal investigator demonstrate your willingness to commit your time to help with their problems and issues. The most important commitment that you can make to your research team is your time. Choose to build rather than simply retreat to a dream of a good research endeavor.

There are two purposes for this effort. One that pays an immediate dividend is learning what it is that each of your team members needs in order to carry out their component of the research project. Changes in technology may require new equipment. A computer may need a replacement hard drive or a new peripheral device. There may be a personal issue that requires not your solution, but your sympathetic ear. Although one might argue that theoretically, channels already exist to handle these needs, in reality, the utility of those channels are amplified by your attendance. You're perceived as choosing to take the time away from something else "more important" and instead spend that time with someone who may not feel they should have access to your time. This effort on your part speaks volumes about your commitment to the project.

> **The most important commitment that you can make to your research team is your time.**

Years ago, I was responsible for a large grant involving heart disease. On my project team, I had a project manager whose job it was to help coordinate the activities of a team of coinvestigators located at other sites. The encounters between the two of us were abrasive. He would frequently confront me with a new problem that I didn't know the project had, demanding an immediate solution. This individual was not a bad person, and actually, he was a very good project coordinator. Unfortunately, our interactions seem to always take place in crises settings—settings that became increasingly intemperate and unhealthy. While I did not wish to avoid the newly discovered problems with the project, I certainly wanted to stay away from these tempestuous encounters.

In an act of desperation, I hit upon the idea of meeting with the project manager on a regular basis, whether we had a new problem or not. He and I began daily half-hour meetings, which, once started, spanned five years. As we spent regular time together, I provided for him what he needed from me as his supervisor—access. From these meetings, I learned much about the basic operation of the project. In addition, we both learned each other's operational styles. Thus, we developed a joint management operation that served us well. What I was required to bring to this effort was my time.

I must confess that as I considered instigating these meetings, part of me rebelled against the very concept of scheduled encounters. That part of me contemptuously complained that here was yet one more piece of my precious time that would be sliced out of my day. Nevertheless, all that was required of me was the time sacrifice demanded of an investigator. Bluntly, the time was coming out of my days one way or another. It was up to me whether I supplied that time constructively in meetings or destructively in acrimonious conversation.

Also, it might be useful to note that the real solution to this dilemma was not the decision to begin these constructive meetings. The actual solution to this problem was my reaction to the discontent of the disruptive confrontations. Essentially, choosing to take the time to consider the problem led to its solution and the resulting meetings. Again, the solution was time commitment.

The coordinated application of attitude, time, and knowledge as you get to know your people can strengthen your project in several dimensions. Attitude simply means that you choose to offer the best of yourself to help each member of your team accomplish their tasks. Be open to the concept that revealing your best outlook is as important a contributor to the team-building concept as your time commitment.

An investigator who spends time with an individual on his team when he doesn't want to meet with them merely means that he has chosen to take the time to inflict his team member with his discourtesy and bad mindset. It is a peculiar facet of human nature that each of us can sense when the person we are talking to or interacting with does not want to be with us. This is communicated through a medium of emotions that is only partially conveyed with words.

In one project, the junior principal investigator obtained a grant to develop a research team that would work on a project that spanned several years. She assiduously assembled a collection of investigators, nurses, and programmers. The nature of the project was such that these research personnel worked in several cities, and therefore, the only way to conduct business on a regular basis was by conference call. In the beginning of the research project, this principal investigator was communicative and open. Her willingness to admit an error in judgment and to accept the correction of her team members not only generated an atmosphere that was a pleasure to work in, but also created a fertile intellectual environment that produced good scientific fruit.

However, over time, the principal investigator's attitude changed. Before, the principal investigator was pleased to engage in discussions, answer questions, and guide the discussions of his team. Now, she was clipped and sharp in her comments and seemed eager to bring conversations to a premature end in order to get on with her daily agenda. Her tone became increasingly clinical, cynical, and critical.

Another change in her style was in how she used her information base. This principal investigator had always been completely knowledgeable about the science of the research. In the beginning, she used this fine familiarity to facilitate discussions among the research team. Now, it was used to bluntly reveal the weaknesses of the arguments of her fellow investigators and staff, that is, as a weapon to browbeat the staff. Her repository of knowledge, although utilized for the research project, was used against her colleagues.

The reaction of the research team was predictable. Growing tired of the blunt and condemnatory comments of the principal investigator, they volunteered less and less during the conference calls. Their silence shouted volumes about the attitude of the PI. One by one, the team members began to withdraw the very best of their talents, insight, and expertise from the research project. It would not be fair to characterize the staff as embittered; however, just like we learn to avoid exposing our skin to acid, they chose to avoid the caustic comments of the principal investigator. Although the research project produced a few manuscripts, its counterproductive atmosphere of unpleasantness led to the project's rapid productivity demise. The more strident the principal investigator became, the less product the research effort generated.

As a junior investigator, recognize that you have and can succumb to personal tendencies that would lead to this disparaging management style. However, let those dark inclinations be consistently overcome by your willingness and capacity to become an affirmative, encouraging, and abetting leader. Bringing the best attitude to a research team encourages the team members to bring their enlightened best to your research project.

While there are some workers who have the unique and special personalities that allow them to give themselves fully even though they receive little respect in return, many more workers require at least a modicum of respect and dignity if they are to redouble their efforts on your project in a crisis.

It is sometimes helpful to recall that you're never more than a single telephone call or email from humility. The progress of your grant can be jeopardized by unseen difficulties, and during these times of trouble, every member of your research team will need to pitch it. Specifically, they will need to work harder than they have had to or may expect to for the good of the project. That decision that they make to go the extra mile is made up of both professionalism and goodwill.

A good way to begin to know your team members is to learn from them their backgrounds and trainings. Get to know what they like about their jobs and those professional activities that they must engage in but don't care for. As time progresses, you'll also come to understand their strengths and weaknesses. The respect that you gain from them places you in the unique position of actually helping them to strengthen their weaknesses. This mutual strengthening activity can be the most important contributions that you can make to your colleagues. Not only do they benefit from this calibration and growth, but they also strengthen the research project in the long run.

One way or the other, your work with your team will leave an impact on them. Whether you ride them to the ground or build them up, you'll instill something. Sacrifice your time to build and develop your team, then let you and your team build the research project.

Learning to Lose

A trial attorney is not a good trial attorney until he has lost a case.

—A lawyer's adage

You cannot be the best leader until you have dealt with and absorbed defeat. In science, defeat comes in many guises. A major grant may not have been funded. A manuscript on which you're first author has received nothing but a steady stream of rejections from the peer-reviewed literature. Another research team may reach a goal that you had hoped your team would reach first. Your fund-raising effort may have fallen far short of its goal. An interview by the media may have become a public relations disaster. These types of setbacks lead to loss of stature, loss of a promotion, loss of prestige, perhaps ultimately, the loss of your job.

There is perhaps no better time for the team leader to set an example than in the face of failure. As the leader, all eyes will be on you to determine how your team will conduct itself in its moment of crisis. Thus, how you choose to carry yourself in this critical time is central to the reaction of your team.

In the face of failure, there are really only two types of behavior in which you can engage as team leader: character destruction or character construction. Your choice is based on your source of self-worth. The performance-significance point of view (i.e., the perspective that permits your sense of self-value to float up or down based on your performance) fails miserably here.

By linking your own sense of value to the success or failure of your operation, a spectacular failure will dictate that you're of less worth due to the inadequate performance of your team. Demanding superior judgment from yourself in a time of crisis can be a near-impossible expectation when you have permitted your sense of self-worth to be diminished, and your character to be reduced, by the team failure over which you have presided. While you may not be emotionally paralyzed by the loss of self-worth, its loss complicates your ability to perform in a new and complex environment. Reduced self-worth can produce in you an awkward and unbalanced approach to the new and complicated environment of failure. This deep loss that you feel can be quickly transmitted to your team.

Defeat hurts. However, there is a difference between the initial painful reaction to a loss and the creation of long-term personal damage. Defeat can and will break your heart. You must not let it break your spirit.

By patiently building up a sense of self-worth that is independent of your activities and performance, you may rely on that undiminished sense of value to power your actions during this time of crisis. You're able to retain your judgment, your standard, your intellectual prowess, and your ability to appraise a situation during these threatening days. With no fear of loss of self-value, you're not threatened in an important way and face no lasting damage from the defeat. This fundamental understanding of the tightly limited impact that defeat has on you personally permits you to be confident and reassured. Your confidence is easily sensed by your team members and by outsiders as well. With this spirit of assuredness, your team is free to apply the best of its talents to learning the major lessons from its defeat.

Becoming a good leader involves the ability to accept and absorb the right lessons from defeat. Specifically, these right lessons are only those lessons that lead to the growth of your character and the development of the character of your team. This automatically relegates blame and condemnation to conduct unworthy of a fine team leader. As the leader, you should first take responsibility for the defeat and then focus the attention of your team on rectifying the defect that led to the defeat. Don't fix the blame—fix the problem. As a leader, times of defeat should be among your best times. During this difficult period for your team, their confidence should be built up and not destroyed. Your team will rely on you for this.

Being a leader makes you the "point person" for criticisms directed at your team. Sometimes this effort leads to calls from others (and occasionally, from within your team itself) that you be replaced. As was the case for facing other crises of leadership, your sense of security is the key to confronting challenges to your authority. Look at these engagements as an opportunity for your character to grow and strengthen rather than as a confrontation in which you'll be damaged. Remaining secure in your own sense of value allows you to deal with threats to your authority on an even footing. This "self-ownership" ensures that no core damage will be done to you, regardless of the outcome of the investigation. Being immune from damage, you're free to state your position clearly, directly, and most importantly, in a nonthreatening way. This level of security allows you to defend your decisions, not as someone who is besieged, but as a confident scientist who has learned from her mistakes and is ready for future challenges. As before, it sometimes pays to remember that what can be taken from you, in the end, may not be worth having.

If someone threatens your authority, recognize that threatening your authority is not threatening you. All of your activities should have as a consequence the growth of your own character or the development of the character of your team.

Final Comments on Leadership

Locked in the heart of every competent junior scientist is the spirit of a leader. You'll have to work hard to first find and then develop these skills, but as a junior investigator, your first task is to acknowledge with your own approval that you can become a scientific leader.

The key to successful leadership is security and a solid sense of self-valuation independent of your outward experiences. With this core in place, you're protected from the personal damage that comes from the demands of leadership. Moreover, you're free to develop and use the best of your intuition, knowledge, training, experience, and expertise without the fear of damage to you. You're free to give the best to your team because you fear no damage from the consequences of your best-considered actions. You can avoid the unfortunate habit pattern of being at your best when you're liked. Additionally, you cannot take care of those for whom you're responsible if you're preoccupied with your own sensitivities. If you can win this fight for your own self-worth, then many more victories await you.

Becoming a leader is a gradual, stepwise process. You gain experience, making the small predictable and unavoidable mistakes of someone growing. Start with some small and limited opportunities for leadership. Seize the opportunity to chair a subcommittee or a grant review session. This allows you to make mistakes from experience without paying too great a price for them. You're going to make some mistakes; be sure to learn the right character-constructing lesson from them.

Build your research team carefully. If your leadership is fair-minded, diligent, and capable, then the search for competent colleagues to join your research effort will not be in vain. Once you have chosen and built your team, resist the desire to do all the team's work in person. This leads to overexcited, zealous, and ultimately exhausting and counterproductive efforts on your part. Don't overdirect. A good team, like a good string instrument, requires not a strike but simply a touch to make it vibrate with music.

Trust competence, encourage initiative, and do not impede ability by needless direction. You have assembled your group based on their experiences. Give them the opportunity to use and develop their expertise. Just as you'll make mistakes as a leader, so they will make mistakes as members of your team. Know the qualities and limitations of your subordinates. Challenge and encourage them to grow.

Also, as leader, when it is time to decide, make a decision. Having prepared thoughtfully and carefully, make the decision that must be made. While one enemy is rashness, a second and equally dangerous adversary for a leader is irresolution. Many scientists who disappoint their own expectations do so because at the controversial moment, they will not accept responsibility for an immediate decision that they are called upon to make.

Finally, remember that everyone is a subordinate and a superior. Don't be a generous and kindly leader, but a difficult and touchy group member.

References

1   Peter L. J., and Hall, R. (1969). *The Peter Principle: Why Things Always Go Wrong.* William Morrow & Company, Inc. New York.

21

Presenting

Introduction

There can be no doubt that as a scientist, you'll have to make presentations before audiences. These lectures will be to colleagues, coworkers, and sometimes competitors. You'll make presentations to both advocates and to skeptics. In each of these settings, as the presenter, you'll have to publicly expose yourself to the comments and criticisms of others.

To be an effective speaker, the ability to communicate openly and clearly with strangers in public requires uncommon strength. While there are "natural speakers," the majority of us must work at developing this skill. This chapter's goal is to show that you have the internal strength that you need to be an effective and persuasive public speaker and to provide guidelines on how to harness that strength.

Sources of Information

There is a wealth of information available about the art of giving presentations. Among the most helpful are as follows:

http://www.public-speaking.org/public-speaking-stagefright-article.htm
http://www.brookes.ac.uk/student/services/health/presentation.
htPresentation Guidelines 2004.pdf
http://clubs.mba.wfu.edu/speakindeacons/Resources1/Ways to Control
Presentation Anxiety.doc

http://www.aacc.cc.md.us/com111/mod3.htm
http://totalcommunicator.com/jitters_article.html
http://www.bradford.ac.uk/acad/civeng/skills/pubspeak.htm

This is by no means an exhaustive list. These websites provide useful, clearly understood discussions of the fear most people have of giving lectures. In addition, they offer useful mechanisms in helping to plan for a presentation. Some of the discussions in this chapter will draw on the content of these resources.

The Rise and Fall of Stage Fright

Any honest discussion of the art of giving presentations must acknowledge that most people rebel at the idea of public speaking. In an often-quoted news article, the *London Times* reported a study in which three thousand interviewed subjects were asked to list their greatest fears.[*] Many concerns were easily anticipated, for example, fear of flying, fear of heights, fear of disease, as well as fear of animals, insects, and loneliness. However, 41 percent of the respondents reported that they feared giving a public presentation above any other. They actually rated this fear greater than their fear of death!

The Reasons for Presentation Anxiety

While it's hard to envision that people are better prepared to part with their lives than to ascend a podium, clearly many people have a deep and sincere dread of speaking in public. In fact, the anxiety leading up to the experience is as bad as making the presentation itself. Each of us has experienced part of the constellation of symptoms association with presentation anxiety, more traditionally known as stage fright. This complex includes blushing, sweating, shaking, rapid speech, and stuttering. In some people, stage fright produces mental confusion in combination with incomprehensible speech.

These symptoms are a response to a physical fear reaction. The adrenal glands, in concert with other hormonal symptoms, react to our fearful response. This reaction of our body is as though the threat felt by us is physical, not emotional. Responding to this perceived physical threat, epinephrine, norepinephrine, and other hormones are released into our systems, priming us to prepare for a powerful attack. Our heart

---

[*] Taken from http://www.aacc.cc.md.us/com111/mod3.htm.

rate increases and our blood pressure rises as the cardiovascular system prepares for a physical danger that is actually not present. Blood rushes from our gastrointestinal tract to our muscles, giving us the sense that "butterflies" are in our stomachs. Our breathing rate increases, and our palms sweat.

This is a strong concerted reaction to a material menace. Your body is preparing you to survive a physical assault. But of course, there will be no attack, and there is no physical threat. There is only the fear. Stage fright is the emotional complex that is triggered by fear.

Of what is the speaker so afraid? There are many opinions about this, and most likely, each of them is right in one circumstance or another. The common explanations for this fear is that the speaker is in a new and unfamiliar situation. Another is that the speaker is fearful of being the sole focus of an audience's attention. This is commonly linked with a fear of isolation. As the speaker, you and you alone are the cynosure— the center of attraction. With the observers concentrating on you, the fear may be that they will concentrate on your appearance, your diction, and your idiosyncrasies. This can easily turn into the fear of being ridiculed. This concern goes together with the fear of looking foolish, of mental befuddlement, of speaking irrationally, of losing coherence. The anticipation of these joint outcomes can overwhelm the speaker before he actually speaks, producing the neurohormonal changes discussed earlier. Ergo, stage fright.

The Response of the Audience

Ultimately, the anxiety-struck speaker fears being hurt and sustaining damage by giving the lecture. But an honest and fair appraisal of this fear reaction requires us to ask how likely is it that such damage will be sustained. An amazing fact about presentation anxiety is that despite all the fears that may consume the speaker, her audience usually don't receive the signs of nervousness that she fears are being transmitted.[*] In fact, each individual in the audience has a set of observations and a personal set of self-absorptions that keeps them from focusing solely on the speaker.

Consider your own reactions as an audience member. Aren't you too busy fumbling with the agenda, or speaking to a colleague, or turning your cell phone off to pay close attention to the speaker's mannerisms for very long? Since the audience is not concentrating on every detail and facet on the speaker's countenance, its members miss the outward signs

---

[*] http://www.aacc.cc.md.us/com111/mod3.htm.

of her fear. In addition, since the audience doesn't know the speaker well, they cannot detect the subtle signals of nervousness that the speaker fears are being broadcast by her mannerisms and nonverbal cues.

> **At its root, stage fright is a very real fear of an unreal threat.**

Actually, as the speaker, your situation is better than that. The members of the audience share your deep-rooted fear of making a presentation. They know how they would feel if the roles were reversed and they were on the stage. In fact, without saying a word, your simple courageous act of walking before them to the podium engenders a sense of admiration in them. The audience hopes that you'll be able to successfully complete your talk. More likely than not, they are rooting for you, and they do not doubt your ability to complete the presentation successfully. In this regard, this audience's optimism is closer to the truth than the low self-image of the panic-stricken speaker.

At its root, then, stage fright is simply a very real fear of a very unreal threat.

The Hunkpapa Sioux

The Hunkpapa is a tribe of Indians that are part of the Sioux Nation.* This nation, along with the Oglala Sioux, Lakota Sioux, and Blackfoot Sioux, inhabited and freely roamed the great plains of the United States up through the midnineteenth century. Among their leaders and fighters were numbered Crazy Horse and Sitting Bull. Although the Sioux had produced many responsible and compassionate leaders, these two men were held in the highest regard by their countrymen. Crazy Horse and Sitting Bull possessed a special, unique potency. They created and exuded a strong sense of self-identity and purpose, powered by a palpable and irresistible force. The Hunkpapa said that these men "owned themselves."

You can confront the challenge of presentation anxiety by resolving the issue of ownership. Overtly, consciously, and affirmatively decide who owns you, then act on that decision. If you reclaim ownership of yourself, then stage fright is dealt with through disassembly. It is a simple replacement process. Presentation anxiety works by filling you with self-destructive images and thoughts, leading to self-destructive reactions.

---

* *Sioux* is an adaptation of the French word for "enemy."

When you actively and affirmatively replace this complex with positive deep-seated beliefs in your own merit, you recover your sense of value and purpose, feel no fear (since you're not threatened), and the reaction to fear never ignites.

Specifically, confront anxiety by reasserting your personal sovereignty. Reacquaint yourself with what you stand for. Establish again in your own heart and mind your irreplaceable value as the individual scientist that you are. Anxiety is a fear of chimera. The reality is that there is only one person, and throughout the existence of humanity, there has been only one person with the unique combination of abilities, strengths, knowledge, curiosity, diligence, and intuition that you have.

You're that person, and the installation of these capabilities and insights within you first imbues and then empowers you with value separate and apart from external concerns and performance. Even if your presentation was an absolute and complete failure, its failure does not diminish the value of your talents and must therefore not reduce your sense of self-value. Your sense of merit can never be taken—it can only be given away.

Retain and strengthen ownership of yourself. Stage fright is an attempt to rob you of your sense of purpose and self-worth. Observe, however, that the presentation and its associated stress will quickly pass away. Any audience reaction to it, good or bad, will rapidly go by the wayside as well. Your personal sovereignty—that is, your appreciation and approval of the value that your unique personality and combination of talents and abilities represents—does not pass; it is constant. They are there for you to appreciate, to draw strength from, and to develop. Presentations are ephemeral. Personal sovereignty is solid and steadfast. Recovering and reasserting it are central.

Preparation

There is no substitute for knowing the subject matter for a talk. The positive impact of the previous discussion about personal sovereignty is minimized if the ultimate source of your discomfort as a speaker is that you simply do not know the material about which you're going to talk.

One common reason for inadequate groundwork for a lecture is that the speaker does not set aside enough time for this preparation process. It is important not to underestimate the amount of time this process will consume. The rule of thumb that I use is to first quickly estimate how long it will take me to prepare for the talk and then simply double that estimate.

Adequate preparation is its own reward. It bears much greater fruit than just the mere product of rote practice. Preparing for a talk is the opportunity for you to review your own work and to find the best perspective for its inclusion into the body of scientific knowledge. This effort is not without its own sense of well-deserved satisfaction to which you're entitled.

During your preparation, be sure to not just reference, but also to understand the work of others. If this work appears in the literature, refamiliarize yourself with the material's contents. This can be invaluable, and not just for your own reeducation. At the time of your presentation, the colleague who made an important contribution to the field may be in the audience. If this individual chooses to ask a question, being cognizant of his contribution places you in a perfect position to construct a responsive answer.

If any of your presented material contains mathematics or arithmetic computations, be sure to check the calculations again. Stop at different points, and recheck your assumptions. Do the calculations in a sequence that is different than the sequence in which you actually carried out the computations. For example, if you must compute the formula $A(B - CD)$, then check your answer by calculating the equivalent $AB - ACD$. It is far too easy to have "your eyes disconnected from your brain" when you review your own mathematics. Checking in a different way is refreshing, ensuring that your calculating faculties are fully engaged in the review process. However, all this takes time, so find the time to do it.

The Layman Rule

The audience of your talk is hoping and expecting to be educated by you. Therefore, although you may have a good deal of information to share with them, you must focus on conveying that information effectively. Make your presentation lucid and digestible. You can learn a great deal from the audience's response to the presentation that you make. Their critical appraisal of your work can sharpen your own scientific focus. In fact, you may get a new research idea based on a discerning comment by a colleague in the audience. However, for these criticisms and comments to be helpful, your colleagues must first understand your presentation.

In my personal ranking of criticisms that my presentations receive, I do not mind hearing, "I know what you're saying, and I think that you're wrong." This is a comment that leads to debate, the outcome of which is good for me and other members of the audience. The worst criticism that I can receive is "Lem, I have no idea what you just spoke about!" This indicates that I have wasted the time of my colleagues as they try to understand an incomprehensible presentation of mine. It is this criticism that I work hardest to avoid.

With this foremost in my mind during my preparation, I try to keep the points of my presentation as clear and as elementary as possible. I call this the "layman rule." My goal is to make a presentation clear enough that a literate layman can understand it.

This rule's purpose is not to insult the audience; its execution does not imply that the audience is "no smarter than a layman." Patronizing the audience is an important mistake to be avoided. The layman rule is instead based on the fact that the audience does not know my work as well as I do. Some components of my discussion will be new and unfamiliar to them. Sometimes, the concepts on which the talk focuses are ideas with which the audience is familiar, but they are just not used to having them expressed in the way that I have chosen. It takes the audience time to understand my perspective on the scientific question at hand. As the speaker, it is my job to keep the audience together and "with me" rather than to "lose them." By presenting the material with the same clarity that I would for a layman (i.e., a literate nonspecialist), I hope to make the talk transparent to everyone.

If you accept the concept that the speaker should make her presentation as clear and as lucid as possible, it then makes sense to apply the following guidelines. Begin your talk by telling your audience what to expect from you. End with a clean, crisp summary. Emphasize the important points and deemphasize the unimportant ones. If you're using variables, explain each one clearly. If you portray a graph, explicitly take a moment to describe the definitions of each of the axes. Avoid complicated tables and multicolored graphs unless you have a great deal of time to dwell on them and explain them. If you're saying something new that the audience has not heard before, then give them the opportunity to digest what you have explained before moving on.

If you're using slides, design them as simply as possible. Many times, I have been in the audience of a talk only to be frustrated by the presenter who announces, "I know that you can't see all the details on this next slide, but . . ." Why has the presenter inflicted this slide on his audience if he knows the audience can't see and understand it? Is this talk for the audience or not? Your goal is to keep your talk as simple as possible so that the ideas that you articulate are clearly communicated with minimal distraction. Respect your audience [1].

Another advantage of the "layman rule" is that its invocation challenges my understanding of the material. After all, if I can't explain the concepts presented in my talk at an elementary level so that an attentive layman understands them, then perhaps I don't understand the material as completely as I should. In this case, I have more work to do.

Walking the Plank

Rehearsing your talk is time-consuming, but of great value. A practice session forces you to concentrate on the small but necessary details of the talk. This microfocus on the minute issues of your presentation is necessary to ensure the accuracy of the factual content of the lecture. However, the entire talk has to fit together as a complete mosaic. A practice session helps you to integrate your talk and see how well its component pieces fit together. The sections of the presentation must support each other. Its flow must be both logical and natural so that the audience is not confused or distracted. These features are nonscientific but aid and support the comprehension of the audience. Finally, the talk must be a certain length. It is hard to have a precise estimate of the duration of your talk without actually giving it.

The second reason for rehearsals is that they give you the opportunity to practice controlling your own nervous energies. Planning the presentation is one thing. Actually talking through it is quite another. Consider the following analogy.

Imagine a wooden plank, ten feet long and one foot wide. It is lying flat on the ground. When asked, almost everyone can walk that plank without falling off to one side or the other. Most people can walk the plank effortlessly, without any thought, particular exertion, or special skill.

Now, take that same plank and raise it until it is one foot off the ground. At this height, fewer people can walk it without falling. If you raise the plank until it is ten feet above the ground, you'll find that most individuals are not even willing to try to walk it. Only the rare individual can walk the plank elevated at one hundred feet.*

The proportion of people who can walk the plank shrinks rapidly as the altitude of the plank increases. Yet the required skill to walk a plank

---

* Taken from comments by Gen. George Thomas, 1863.

ten feet long and one foot wide remains the same. What has changed is the required confidence in your capability to carry out what remains an easy task. Learning to trust in your own abilities, regardless of the external consequences, takes time and patience. This is what is gained in rehearsal. The more important the presentation (i.e., the higher the plank will be set in the end), the more important the need for rehearsal.

The ease of recording your performance on videotape can amplify the ability of rehearsals to improve your confidence. First, the use of videotape adds to the realism of your practice presentation. Secondly, watching the videotape gives you the opportunity to be in the audience of your own practice presentation. This perspective will permit you to make some countenance observations. Do you speak too fast? Do you speak in a monotone? Do you look down too much, paying no attention to your audience? Do you move your arms too much, distracting the audience? Do you stand like a statue? These mannerisms are clearly revealed in a videotape of your practice presentation.

Continuing with this issue, you can use the video rehearsal to aid in changing any of your bad habits that appear during a presentation. For example, I found that when left to my own natural tendencies, I speak too fast. While I wanted to correct this, moving to the other extreme would be intolerable. I needed a way to calibrate my speaking speed. By using a videotape to tape, then view, then retape, then review a short reading session of mine, I was able to find the best speaking speed and practice it until it became effortless.

Seventy-two Hours to Go . . .

Every scientist's last-minute preparations for a major presentation are different. I am best served by the following. Three days before the actual presentation, I bring my content preparations to an end. I have completed my reviews. I have prepared my slides. Rehearsals have come and gone. I know where I am to give the talk and how to get there. I have a reliable estimate of the audience size. I have found the audiovisual equipment that will be used to support my presentation and have tried it out, assuring myself that I am comfortable with its basic use. If I need to, I can give the talk without much in the way of audiovisual support. All this planning work has been completed. Now, three days before the scheduled presentation, this preparatory component ends, and a new readiness phase begins.

During this second final phase, I consciously and deliberately remove my focus from the science and reseat it on myself. Specifically, I draw on my character, recover, strengthen, and reassert ownership over my life and its guiding principles. I overtly review what my career stands for. I reflect on the better natures of my personality. I recount the times that in a moment of despair, a family member, colleague, friend, or senior scientist said exactly what I needed to hear, the way I needed to hear it.

During this period, I shun new manuscripts, avoid studying email, and distance myself from last-minute advice. Instead, I insist that I get adequate sleep and rest. I watch a favorite movie or read a chapter from a beloved book. Having already mastered and controlled the science in my presentation, I now choose to reassert mastery over myself and enjoy some of life in the process.

Now, being centered and secure and in full recognition that my purpose, value, and merit are unaffected by my actual presentation or any reaction to it, I am prepared to speak to my audience.

Reverse Radar

Sometimes in preparation for a pivotal lecture, it is easy to be plagued by new distractions. These distractions come disguised as questions that have no good answers. Examples are "What if the president of the company comes to my presentation when I am discussing the most difficult part of the talk?" "What if no one chooses to attend the session?" "What if someone interrupts my presentation, consumed with violent disagreement?" "What if they laugh at me?" These types of tormenting questions can be gut-wrenching and interminable. If you're not careful, these Lilliputian concerns can tie you down.

Years ago, in a moment of frustration, I decided to put these questions to the test. Specifically, before a presentation, I would make note of these self-afflicting questions (a warning that invariably involved someone else's action and not mine). After the presentation was complete, I would take a moment to review whether the warning came true. In no case did the event that was predicted by these admonitions come to pass! In every example, the issue raised by each of the "what if" questions utterly failed in its prediction.

Of course, unplanned events do occur during presentations. Slide projector bulbs can and do go out. Hard drive performances become erratic. Media storage devices are incompatible with each other.

Operating systems still crash. These unfortunate events do occur.* However, the key observation is that the predicted events do not. Thus, as a practical matter, these early monitories are not really warnings at all; they are merely distractions from what actually will occur. By placing your focus on a false prediction that is most commonly the product of anxiety, you can be caught unprepared by the occurrence of an unpredicted event.

In the face of these observations, I decided that it would not be sufficient for me to ignore the "what-ifs." Instead, I have gained assurance that what they warn of what will not occur. Since they have always been false, they can be treated as a "reverse radar," not pointing toward the direction of danger, but instead pointing away from it. Thus, a pre-presentation "thought alarm" attempting to warn (or frighten) me about the danger of a question that would be put to me about whether I had read an obscure manuscript years ago can be treated as an alarm about a question that will do me no harm. I can therefore discount it and return my focus to more pressing and helpful considerations.

Two Minutes to Go . . .

There are only a few minutes to go before your talk. The prior speaker is concluding her presentation. This is the most agonizing time. Commonly, after my labors and preparations, I find that my pulse has nevertheless inched perceptibly upward. However, I know that in less than five minutes, it will be back down to its regular, steady, and reliable rhythm. I therefore take my mind off this useless pulse-monitoring activity and instead choose to place it in the following story:

> A young officer during the US Civil War was given his first assignment. With several hundred soldiers under his new command, he was ordered to take his troops ten miles (half a day's march) and from there to dislodge an enemy force from its position.
>
> This officer, after organizing and informing his troops, started them on the march, himself in the lead. As the march began, this young officer realized

---

* At the beginning of a presentation that I gave at a major corporation, I scanned the audience and identified my college advisor from thirty years before in attendance! While this was not bad, it certainly was an unanticipated surprise. I had no idea that he was no longer in academics.

that although he wanted to be an officer, he had never actually led soldiers into battle before. Even though he knew many of the frequently used unit formations and had memorized many of the orders that commanders commonly gave in battle, he had never given them in a fight. It occurred to him that he was very inexperienced (perhaps too inexperienced, a quiet voice whispered) to be given this command.

As he led his columns of troops over the hilly terrain toward his objective, he reflected on other issues in his life. Aside from his military education, he had not done well in school. In fact, he also performed poorly during his military training. He had developed a taste for alcohol that he had been unable to control, and occasionally, he had to be revived from a drunken stupor by other officers. He had tried to be a farmer once—and had failed.

In fact, he reflected, every business he undertook in order to support his family had ended in failure. Ultimately, he wound up working in his father's tannery before the war because he had failed at everything else. His life, so far, was filled with failure, and now this.

He revived himself from his brooding, facing the recognition that he could not go on. As he and his soldiers approached the last rise between them and the enemy, he opened his mouth to give the order to halt the advance and return to camp.

However, at that moment, when he knew within his heart he was going to fail as a soldier too, he gained his first unobstructed view of the enemy. He saw that they had seen him too and that they were moving rapidly, but moving rapidly *away* from him. As he sat in amazement, watching the last of his adversaries disappear from sight, it occurred to him that they had been just as afraid of him as he was of them. This was a turn of the question that he had never considered before and did not forget thereafter.

The young officer was Ulysses Grant.

Giving the Presentation

When it is time for you to give your presentation, you simply need the strength to follow through on your commitment to yourself. Don't think—simply do what you planned to do.

Countenance

When I first started speaking publicly, I used to spend time memorizing the first one or two minutes of my planned remarks to the audience. This maneuver helped me to relax and to settle down. However, although it was a useful exercise, I no longer use it because it inappropriately changed my concentration. I am not well focused on my audience during this recitation but am instead focused on steadying myself. Thus, I now make my final settling adjustments before I speak to the audience, helping to ensure that when I begin speaking, the audience has my full attention.

An important key to relaxing your audience to what you have to say is your use of vocalisms. Vocalisms are not the words, but the sounds that accompany your speech. Specifically, they are the tones, pronunciations, speech rates, pitch, and inflections in your voice. The philosopher Friedrich Nietzsche thought of vocalisms as the most intelligible part of language. When you're confident, your voice has one sound. When you're hesitant, it has another. It has been said that just as lyrics need music to come alive, spoken words need vocalisms.

The challenge for the speaker is to produce the right vocalisms in order to engage their audience. To address this, consider that just as vocalisms are the music behind the words, there is emotion behind the vocalisms and a person behind the emotion. Therefore, to correctly influence vocalisms, the speaker has to be the right person, that is, has to have the right attitude. From the right attitude flows the right emotion, and from the right emotion comes the right combination of vocalisms.

The right attitude is commonly an open one. In order to open up your audience so that they can hear what you have to say, you must open yourself. When you relax and open up, your audience follows suit.

Actually, a very simple process can produce the timely transformation that you need. As I approach the podium then turn and face the audience, I ask myself how I would explain the material that I am presenting to my most special loved one. How would I educate, convince, and persuade her of the content and implications of my work? Immersing myself in this context produces the change in my attitude and outlook that I need. All sense of anxiety, of resentment, of frustration, and of anger drain away. In full anticipation of being trusted, and with the certainty of forgiveness for any mistake that I might make, I cannot help but relax. Relaxing, in turn, reconnects me to my best faculties. My memories, my vocabulary, my sense of humor are all once again under my control and command; and I can enjoy the experience. My vocalisms adjust to my attitude and emotions, and I have created the environment in which I have the best opportunity to reach my audience.

Actively choosing to submerge myself in this mindset evokes from me the perfect countenance. When as the speaker I relax, the audience members themselves respond by relaxing. When the audience is relaxed, they more easily learn from my presentation. The rest is anticlimactic. The presentation proceeds easily and is over much too quickly.

Slideshows

Although I am compelled to use slideshows in my discussion, I do my best not to rely on them. All the practice that I have engaged in during the early preparation for the talk permits me to essentially give the lecture without the slides. If I allow myself to rely on them, then I find that I drift into concentrating on the slides and not the people in the audience.

The appropriate focus is typically the result of a balancing act. It is, of course, useful to have the slides available to you. Certainly, in a technical talk, you don't want to have to force yourself to memorize many small but important details that you'll provide in the lecture. Also, if you get distracted, for example, by an unanticipated question that temporarily interrupts your train of thought, the slide is there to help refocus you. However, by simply reading your slides to the audience, you lose a potentially important connection with the audience.

My goal in giving a talk is not reading my slides, but reading my audience. I want to gauge their reactions to my talk. Are they upset? Bored? Energized? Focused? My sense of how the audience is reacting helps to guide the words that I use to reach them. I may need to alter my voice cadence or volume. I may make a point that I had not planned to make, or I may pull back and not extend myself as far as I had planned. Staying and remaining calm is the key. Focus on the audience, and discover what moves them.

When You Make a Mistake

Eventually, you'll make a mistake during a presentation. You'll forget a detail. You may garble your description of material covered by one slide.

You may be sternly corrected by an audience member who has identified an inaccuracy in your presentation. There are two useful steps that you can take to handle this type of occurrence.

The first is to anticipate them. You're delivering a presentation to an intelligent, attentive, and discerning audience. Anticipate that they will find something critical to say about your lecture and that they may interrupt you to make their point. Be prepared for it if and when it occurs. Similarly, expect that you'll make a mistake working through a slide. Although you hope to complete the presentation perfectly and without error, expect that you might misstep. Don't be surprised if despite your best effort, you "blow a slide."

Secondly, keep in mind that mistakes are common and everyone makes them. What distinguishes you is not the error, but your response to it. React by quickly correcting the mistake. One of the best ways to do this is repeat your review of the slide that you garbled, this time saying it correctly, then pausing to make sure that everyone understands the correction before you move on. For example, during a presentation, I said,

> In this sample-size estimate, the use of the dependency parameter decreased the total sample size of the trial by increasing the sample size for the analysis of the primary endpoint that had the largest variance of the estimator.

However, this was a wrong statement that I realized a moment later. I then tell my audience,

> I need to apologize because I misspoke a moment ago. What I meant to say was "In this sample-size estimate, the use of the dependency parameter decreased the total sample size of the trial by *decreasing* the sample size for the analysis of the primary endpoint that had the largest variance of the estimator." I just want to take a moment to make sure that everyone understands my error and the correction.

Try not to rush through making a correction. Despite your best efforts, the audience may find your presentation difficult enough to follow if it contained no mistakes. Mistakes obstruct their views of your main point. When you make a mistake, go back; correct it for your audience, giving them a moment to collect themselves; and then move them forward through the rest of your talk.

Secondly, after you make the correction, proceed with the remainder of your talk as though the mistake did not occur. Don't let the mistake fatally disrupt your equilibrium. It's been corrected, producing only transient harm.

Even the best of prizefighters, on their way to a victory in a fifteen-round fight, will lose a round. In fact, he may lose more than one round. The eventual victor may even get knocked down during the fight. However, the good boxer keeps his sense of balance, purpose, and courage. Despite these temporary setbacks, he comes back to fight the remaining rounds hard, which was his goal all along.

You expected that mistakes may occur. You have acknowledged and corrected it. All there is to do now is to move on.

Finishing the Talk

As the end of the talk approaches, the speaker commonly and correctly recognizes that they are nearing the end of the experience (or the ordeal). This sometimes leads to a new accelerated speaking pattern as he hurtles to the end of his talk. This is an unfortunate reaction that you should resist.

This last-minute rush can undo the fine first part of the presentation that you completed. By racing through the final component of your talk, the audience quickly senses that you do not want to speak to them anymore. As you detach from them, they can disconnect from you, and some of the more valuable points in the last part of your lecture can be missed. Pace yourself and check your speech pattern as you near the end of your presentation. If you need to, make yourself pause for a moment to recenter yourself before finishing the lecture.

If you have carefully planned and clearly delivered your presentation, you can anticipate that there will be several questions at the conclusion of your talk. Questions are a blessing, not a curse. Their presence often affirms that people understood your talk and that they want to engage you. You can gain and impart important new knowledge in this interchange that follows the lecture. If you can answer dynamically, the question-and-answer session can become a spirited repartee that everyone enjoys.

When answering questions at the conclusion of the talk, be sure to answer the question. Speak in a voice that can clearly be heard. It commonly helps if you can first answer the question directly rather than give an elaborate re-presentation of your work. Consider the following question asked at the conclusion of a statistical talk that I gave:

> Is the sponsor of the study, who contributed tens of millions of dollars to this research effort, more likely to be happy with the conclusion of the experiment if they chose to follow your plan?

A common answer might be as follows:

> These research efforts are complicated. They involve many considerations, and many factors have to weighed. Some of these factors were presented in my talk. However, it is possible that the selection of more than one possible prospectively declared endpoint can increase the likelihood that the sponsor will be pleased with the result.

In my view, a better response to the question "Is the sponsor of the study, who contributed tens of millions of dollars to this research effort, more likely to be happy with the conclusion of the experiment if they chose to follow your plan?" would be the following:

> Yes. By prospectively declaring more than one endpoint, and with the a priori allocation of type 1 error to each of the endpoints, the sponsor is more likely to be happier with the outcome, everything else being equal.

Both answers are defensible, but the first is indirect and frankly does more to protect the speaker than it does to educate the audience. It is what I call a "nonresponse response." The second answer is clear and firm. It puts the speaker on the spot, in the sense that he may be asked to defend his assertion. The answer doesn't have to be provocative. But if a question can be answered clearly and directly without being provocative, then answer it clearly. On the other hand, if you're ethically bound not to answer a question, then simply say, "I cannot answer that" and explain why.*

---

* An example would be if you're asked about the results of an experiment that

Additionally, answer all questions respectfully. If a question asked by an audience member appears silly to you, don't let any trace of your attitude enter your voice. In this case, that attitude is likely to be transmitted as arrogance to the audience. If a question is asked angrily, then respond with an answer that is both firm and soft-spoken. When considering your answer to such a question, take a moment to drive fear and anger out of your heart; this will ensure that these emotions do not betray themselves in your voice.

Finally, when you're all done, take a few moments alone to reflect on what you have accomplished. You have just completed an important presentation to a large audience. While you may have felt fear, you were not moved by it. Instead, you remained purposeful, steadfast, and true to your conviction that you had something of value to impart to the audience. You deserve to feel good and thankful for a moment, so be sure to take a moment to do so.

Later, take some time to reflect on the usefulness of the preparations that you made for your lecture. On what should you have spent more time? Were you overprepared on some issues? What maneuver helped you to retain control over your own nervous energies? Did you focus on the audience during the talk? Did you control your speaking cadences? Did you make any mistakes, and if so, did you respond well to them? Did any of the "what-ifs" come to pass? This short solitary review will be helpful as you begin to prepare for the next presentation.

Final Comments

Preparing for a presentation tests your true knowledge of your own work by requiring you to clearly explain it to others. Giving lectures provides a way to disseminate your work. Defending your work in a question-and-answer session can instruct you about the utility of your work and its contribution to science while sharpening your debating skills.

If there is an enemy to be faced in making presentations, it is not the interaction with the audience, but the fear of failure. Don't let presentation anxiety stunt your development as a scientist.

References

1   Feibelman PJ. (1993). *A PhD Is Not Enough! A Guide to Survival in Science.* Cambridge MA. Perseus Books.

---

cannot be released yet because of obligations to your coinvestigators, who have all agreed to announce the results jointly at some time in the future.

22

Junior Faculty Member

Academics 101

If you're entering academia as an instructor or as an assistant professor, then you're crossing over into a world of transformation. While academia retains its distinctiveness from nondidactic institutes or corporations, the academic world is nevertheless undergoing tectonic shifts. While these changes and alterations need not jeopardize your new career, you need to be aware of them.

There was a time when the decision to enter academia involved the consideration of a relatively simple trade-off. Plainly, the researcher had to choose between freedom on the one hand and income on the other. A young scientist entering the academic community retained complete control of his career. He could work on what he wanted, when he wanted. He could speak up in public, virtually unfettered. He could teach almost any course that he wanted. He could grade students any way that he wanted.[*]

What the academician gave up for this freedom was money. In general, the academician's salary was significantly lower than that of the

---

[*] An example of this traditional view in extremis of academic life is the character Indiana Jones (from the movie trilogy) who is an archeology department faculty member at a university in the 1930s. This "academician" teaches the courses he wishes, avoids discussions with students by climbing out of his office window, and leaves the university on a whim to engage in scientific quests. This caricature shows academic freedom at its best and worst.

equally trained professional in the private sector. However, these private sector professionals were commonly compelled to work long hours on scientific projects in which they had no fundamental intellectual interest. They could be peremptorily dismissed from their job. Public comments had to be vetted.

Thus, the choice was simple. While there are always exceptions, in general, by working in academia, you gave up being rich for the ability to be independent. This independence was called "academic freedom."

In our contemporary era, this paradigm has become complicated because the idea of complete freedom has been, for better or for worse, successfully challenged by another idea—accountability. State universities must be accountable to state legislators. Private universities must be accountable to their governing bodies. The caricature of the unproductive tenured professor, drawing salary but generating no intellectual product, has compelled universities to review the productivity of their faculty on a regular basis. Faculty expenditures and positions must be justified, and that justification is obtained through an evaluation of productivity as measured in three major areas: research, service, and teaching. Thus, in the new environment, the academic scientist is never far from justifying her work product to deans and other university administrators.

While this tightening of faculty oversight is not necessarily a bad thing, close supervisory overview of faculty work is somewhat new. In this chapter, we will discuss the modern metric of professional measurement in colleges and universities and then develop a professional framework in which you may carry out relatively untroubled work in a new and turbulent environment.

Defining and Evaluating Productivity

The core of the academic career is scholarship. Academic life is devoted to the pursuit of knowledge for its own sake, free from financial concerns, administrative requirements, and the need to always be right. The academic career is one of two professional classes of activities (religion being the second) where the metric of success has traditionally been cerebral—not material.

By selecting a career in academia, you have chosen to benefit from, as well as contribute to, the creation of an atmosphere where scholarly pursuit and professionalism are ascendant. The three major areas of contribution in academics are teaching, research, and service. However, these contributions are most effectively delivered from an attitude of service, strength, kindness, collegiality, discipline, charity, and leadership.

Teaching

In academia, to teach is to guide the learning of others. Educating students remains a key activity for faculty and is central to the measurement of academic productivity. However, even though teaching remains a core contribution, there typically are few direct financial rewards for teaching. In fact, the dearth of pecuniary prizes for didactic expertise requires college and university faculty to continue to ensure that superior teaching is singled out for special attention and praise. The need for the clear public recognition of these devoted teaching efforts is a perennial challenge before all faculty.

Continuing to place your best effort into teaching in the absence of material or financial awards requires a unique discipline and steadfastness. Typically, the teaching load of a faculty member is measured using a combination of three criteria: (1) the number of courses a faculty member teaches, (2) the number of students in each course, and (3) the percent effort that the faculty member expends in teaching. However, while these simple tabulations provide an evaluation of how much teaching you do, they do not yield an assessment of the effectiveness of your teaching. In order to gain this useful perspective, the students' evaluations of your teaching skills are obtained. This information allows an appraisal of your style, breadth of knowledge, and availability, permitting both you and those who oversee your teaching contribution to assess your ability to reach students.

In addition, your ability to work with students individually is evaluated through an assessment of your willingness and ability to advise students. Serving on MPH or MS thesis committees, qualifying examination committees, and the dissertation committees of DrPH and PhD candidates is an effort that is valued by the university and prized by the student. By working closely with your advisee, you closely inspect and critique their individual research and investigational activities. Observing and then discussing the students' strengths and weaknesses with them provides invaluable commentary for the student to absorb and consider during this formative stage of their scientific development.

Secondly, by choosing to work closely with an individual student, you reveal your own personal study and research approaches to your advisee. Essentially, you open yourself to your student, allowing her to carefully examine your style and technique and permitting her to pick and choose which habits and perspectives of yours she would like to accept as her own.

This intensive process is one that requires not just your valuable time and attention, but also tests strength of character. It takes a secure adult with a good measure of their own self-worth to allow a student to openly make commentary or criticize their approach in a way that they are unaccustomed to hearing. Serving as an advisor requires patience, discernment, and self-discipline. Those faculty who can effectively train and mentor students in this most effective way deserve special consideration for these fine efforts.

Research

A second core contribution that you'll make to academic scholarship is in the area of research. It's called *re*search because rarely is a new idea completely novel in this complex interactive world. The genesis of a scientific concept can most commonly be identified in an examination of the work that has preceded their efforts that therefore must be examined again. In scientific fields where scientific reasoning is commonly hampered by the absence of good knowledge, research and investigation must be a central component of your productivity.

While faculty can generally agree on the need for scholarly investigation, the generation of new knowledge, and the dissemination of that knowledge, there is commonly a healthy and ongoing debate on how research productivity should be measured. What follows is a collection of metrics that are used and a justification of their use.

Publications

The publication of a manuscript is a declaration that the article's contents have reached a standard of quality that is worthy of dissemination to the scientific community. Lists of publications on which you are the author are cited in your résumé. Authors of publications are commonly ranked by their location on the masthead or title area of the manuscript. The value of a manuscript is measured by the quality of the research that generated it, the written description of that effort, an assessment of your role in the entire process, and the standing of the journal in which the manuscript appeared.[*]

---

[*] This is not to say that manuscripts that pass the peer-review test are correct. Sometimes, manuscripts are published because they represent new findings that

The peer-review process is an attempt to ensure that the manuscript's contents are objectively evaluated by outside experts in the field. Since outside reviewers are more likely to be objective than your colleagues or friends, the opinions of these external experts are, in general, more highly valued. The appearance of a manuscript in a peer-review journal is a statement that the research effort and its description has met the current standard of science for dissemination. The more competitive the journal, the greater the number of candidate manuscripts there are that are vying for acceptance, the more grueling and selective the external review, and the more rewarding the final publication.

A second important consideration in the assessment of your scientific contribution to the literature is an evaluation of the role that you played in the research effort. This is difficult to accurately assess; however, the current state of these evaluative efforts focuses on the location of your name on the masthead of a manuscript. If you're in the middle of a long list of authors, then, although the research effort itself is laudatory, it can be difficult to assess your role in the project and almost impossible to view your performance in the development and authorship of the manuscript itself.

Typically, the scientist whose name appears first on the manuscript is the person who is directly responsible for writing the manuscript. Therefore, this "first author" is a scientist who has been intimately involved in the research project. This person provides an outline of the manuscript and may develop writing assignments for the coauthors. The first author writes the sections of the manuscript that can be the most vexing. This scientist takes charge of the development of the bibliography. The first author accepts sections from the other authors, but these contributions must be carefully and critically reviewed before they are included in the paper; it is the task of the first author to ensure that the separately written pieces of the manuscript fit seamlessly together.

When the manuscript must be edited in accordance with the demands of the reviewers and editors, it is the first author who takes responsibility for providing a response to each of their concerns and who also ensures that the manuscript is appropriately revised. The stamp of

---

must be confirmed before they are accepted. The appearance of a manuscript in the literature is a statement that the article has reached a minimum quality and that the results (in the editor's view) should be conveyed to the target audience, not a guarantee that the manuscript's results are correct. See chapter 8 for an example.

the first author's style is on the final manuscript, and its acceptance by a journal is substantial evidence of that author's ability to effectively communicate to the scientific readership. Senior investigators who provide important organizational support and intellectual leadership commonly appear last in the authorship masthead.

While the responsibilities of the first author are generally clear, the specific activities of authors whose names appear after the first author become ambiguous. Therefore, as your name moves down the list of authors, from first to second to third to fourth etc., confusion in the minds of your productivity evaluators about your specific role in writing the manuscript naturally rises. Thus, these manuscripts on which you're not first author carry less-persuasive weight about your ability to write effectively than other manuscripts on which you're the lead author, ceteris paribus.

Of course, articles on which you're not first author but in whose research you have played a major role have important value of a different kind. Specifically, these manuscripts speak to your ability to successfully participate in a collaborative effort. This is itself a critical skill and must not be ignored. However, these articles carry less positive weight in the argument that you can successfully write to the satisfaction of the scientific community.

Moving to the other extreme, published manuscripts on which you're the sole author are of great value since the research is yours and, in addition, you have sole responsibility for describing that research effort. This manuscript's acceptance in a high-quality, competitive journal speaks volumes about your ability to produce solid work and competently describe that work in a way that is deemed worthy of dissemination to the scientific community by others.

Books and book chapters typically carry less weight than peer-reviewed journal articles in science. This is not because these literary contributions require less effort or are less persuasive. The reason that books and book chapters are discounted is primarily because the degree to which they are peer-reviewed is uneven and difficult to judge. Articles and manuscripts that are submitted to peer-reviewed journals are not accepted unless and until the external reviewers and editors have received a text with which they are satisfied. Books, on the other hand, are commonly accepted before they are completed. Many times, books are accepted for publication on the basis of a five-page prospectus (or other writing plan) and one or two sample chapters.

In addition, although the work may undergo a rigorous review upon completion, typically, the author has a good deal more leeway in responding to the reviewers' demands. Specifically, since the book has already been accepted, the authors are not required to satisfy every concern raised by each reviewer. Thus, while the quality of books reaches a standard, and many outstanding texts have been and will continue to be written using this standard, that standard is different from the one used by top-quality peer-reviewed journals. This is one reason why books and book chapters are not commonly authored by junior researchers, but are instead written by more senior scientists who can afford to be less concerned about the hierarchy of productivity measurement.

Judging verbal presentations offer the same difficulty to your evaluator as does assessing the contribution of a book. Making speaking presentations is always demanding, and commonly, a manuscript must be presented to the sponsoring body at or near the time of the presentation. However, it is difficult for an evaluator who has not attended your lecture but only sees its brief description in your résumé to judge the standard to which the presentation rises. While there is a hierarchy of meetings in every field, just as there is a hierarchy of peer-reviewed journals, a presentation is commonly held to be of less value than a peer-reviewed manuscript, everything else being equal.

Grants

Being funded to carry our research is a central component of much academic work. Its major impact on your evaluators stems from the fact that grants (1) demonstrate your intent and ability to carry out research that extends scientific knowledge, the raison d'être of a scientist; (2) demonstrate your ability to successfully compete against other equally capable researchers in the eyes of external reviewers, therefore reaching the standard of an external peer-review test; and (3) the award of a research grant provides funds that benefit your university. While unfunded research meets the first of these criteria and can provide valuable insight leading to peer-reviewed publications, it does not meet these last two tests.

As was the case for the evaluation of publication efforts, there is a hierarchy among grants. Certainly, a grant on which you're the principal investigator presents the strongest case for your ability to describe your research plans cogently and persuasively. In this case, you're the primary person responsible for planning and conducting the science and the main individual in charge of administering the research funds. It is anticipated that you, as the principal investigator, will play the key role in producing the manuscripts that describe the funded research.

The ability and the expectation for you to be the principal investigator on a grant (i.e., "have your own grant") can depend on your field. If you're in research medicine, biology, chemistry, genetics, or epidemiology, for example, you may be expected to obtain your own grant for which numerous granting agencies are available.

Grants on which you're an investigator but not the principal investigator are also important demonstrations of your research expertise. They demonstrate your ability to work as a member of a research team. However, don't miss the opportunity to play a central role in the grant's mission. One such opportunity is to take the lead on a subproject of the grant.

Thus, when it comes to assessing scholarship, manuscripts in the peer-reviewed literature and funded research are among the most influential bodies of evidence. They directly speak to your ability to make scientific contributions that are competitive, well delineated, and relevant.

Community Service

Service is the process by which you place your own needs aside while you use your professional intuition, talents, and skills to provide for the needs of others. While teaching and research are measures of your scientific prowess, service is a measure of your citizenship. There is almost an uncountable number of service needs that you could satisfy. These exist at either the level of the university or the local community, state, national, and international levels.

Your service contributions are certainly desired and appreciated at the university level. The successful and smooth operation of your institution depends on the willingness of faculty to play major roles in the oversight of the school. With the gift of academic freedom, faculty bears the important responsibility of creating an environment of productivity and development. This is an obligation that is successfully met with consistent energy and dedication.

Academic programs are best developed by faculty. Definitions of academic scholarship are best specified by faculty. Junior faculty are best mentored by faculty. The most capable and diligent search for new faculty are executed by faculty. Guidelines for promotion and tenure of faculty are best set by faculty. These responsibilities require an important time commitment from faculty members who commonly already have a full teaching and research agenda. However, the fact that many faculty members devote themselves to these actions speaks to their recognition of the importance of these tasks. Involvement in these activities not only increases your knowledge about the school, but also demonstrates your willingness to put some of your own interests aside for the good of your fellow faculty and students.

The local needs of a community outside the university are varied, allowing you the opportunity to identify an activity in which you have a real interest that also requires your unique combination of understanding and capability. While you may feel somewhat isolated at the university, laboring under the belief that you're working separate and apart from any community activities, it is also true that there are many community groups who themselves feel cut off from the information and support that they need. These organizations have interests or are charged with making decisions that require a scientific base of knowledge. However, they are unsure how to reach out for the required information.

Municipal organizations, health departments, community hospitals, elder care centers, environmental agencies, and secondary school systems are but some of the many neighborhood activities that can avail themselves of your insight and knowledge. Each group commonly has very little in the way of remunerative resources, yet each can have an important need for your help and insight.

Service at the state, national, and international levels provides an opportunity for you to work with other professionals through organizations with a broad reach. This service includes but is not limited to taking part in grant reviews, agreeing to participate in panel discussions, and serving as a referee for a peer-reviewed journal. It also includes serving on oversight boards, writing position papers, and testifying before the legislature or Congress. These actions permit you to work with and learn from senior scientists and researchers while jointly making progress on the issues at hand. You can gain important new experience, insight, and skill by working with senior investigators as you contribute to their work on writing a position paper. This activity can improve your visibility on the state, national, and international level.

Promotion and tenure

In academia, the process of promotion and tenure has both simple and complex components. Every university has its own procedures. Most commonly, this process includes the work of a promotions and tenure committee, whose job begins by gathering relevant information about the candidate faculty member. Determinations are made about each candidate's suitability for promotion and, when appropriate, the awarding of tenure. After these recommendations are reviewed by the school's dean, recommendations about promotion and the award of tenure are then made from the school to the university where they are again reviewed by higher university academic officials, up through and including the university board of regents. Despite changes and adaptations that have occurred over the years, the university tenure and promotion process continue to be one of the most serious and formal activities that these institutions undertake.

Promotion

Promotion is the process by which the candidate progresses through a series of well-defined ranks. Traditionally, these steps have been from the rank of assistant professor to the rank of associate professor and from the rank of associate professor to the rank of full professor. The criterion for these promotions is the candidate faculty member's scholarship record. Thus, the abilities of the faculty member to effectively compete for research dollars and to convert these grant awards into publications that appear in the peer-review literature are central to the consideration of promotion. Also important is the demonstrated capability of the candidate faculty member to be an effective teacher for the university and the candidate's commitment to the scientific community.

> **A junior faculty member who is well grounded, smart, accepts advice and constructive criticism, diligently works in her field, seeks new projects, looks for the opportunity to engage in research, actively conducts that research, persistently writes about that research for the peer-reviewed literature, and teaches will demonstrate that she has the traits worthy of promotion.**

Many rules of thumb have arisen in the hope of providing junior faculty some guidance about the requirements for promotion. One example of such a "rule" is "a junior faculty member must have at least twenty peer-reviewed manuscripts in order to be promoted from assistant professor to associate professor." A second "rule" might be "an associate professor must have a national reputation in their field, and a full professor must have an international reputation."

Such monitories, while informative, must be shunned as hard and fast rules since one need not look far to find counterexamples to their predictive ability. What these guidelines attempt to embody is the sense that a faculty member who is both intelligent and makes sedulous use of her professional energy will demonstrate these traits and characteristics through well-recognized and accepted demonstrations of academic scholarship and research. A junior faculty member who is well grounded, smart, accepts advice and constructive criticism, diligently works in her field, seeks new projects, looks for the opportunity to engage in research, actively conducts that research, persistently writes about that research for the peer-reviewed literature, and teaches will demonstrate that she has the traits worthy of promotion.

Tenure

Since tenure is a measure of the overall value of a faculty member to the university, its definition and criteria for award are less concrete and more nebulous than those for promotion. Traditionally, the award of tenure is the culmination of a long-term commitment made between the faculty member and the university. The university awards tenure if it determines that the candidate faculty member has been and can reasonably be expected to continue to demonstrate a consistent dedication to its mission.

Thus, the faculty member must demonstrate that her activities have been in alignment with the university goals. This includes, but is not limited to, service to fellow faculty, service to the school, service to the university, and service to the broader community. Through awarding tenure, the university hopes to identify that relatively small number of faculty who are willing to put aside their own needs for the good of the institution.

Diligent Days

As junior faculty, the aforementioned metric (or the particular relevant measures at your institution) should be kept in mind. Your goal should be to use your academic freedom to pick the activities that will allow you to engage your insights, strengths, and talents to advance your field of science, thereby earning you fair recognition for your assiduous efforts by your peers.

However, many activities in which you can become involved swirl around you. Research activities may already be available for you to work in, or you may apply for new ones. While you may be handed an initial teaching assignment, once you have begun teaching and become comfortable with the task, handling the predictable problems of students with facility, you can expand (or perhaps contract to a degree) your teaching responsibilities. Service opportunities will grow as well, permitting you to pick what you would like to be involved in at a local, national, and even international level.

This freedom of opportunity provides a unique challenge. The early recognition of this challenge can lead to the creation of a rewarding and sustaining personal work lifestyle that allows you to reach your goal.

Balance and Discipline

As a single person, you learned that you cannot date all the attractive people simply because they're all attractive. You can't choose everyone; you must instead choose someone. Similarly, you cannot be involved in all the available university opportunities in teaching, service, and research merely because they are all appealing. You cannot select everything; you must select from everything. Doing the opposite, and being involved in all activities, produces chaotic days. The fundamental difficulty presented by these inchoate days is that you cannot bring the best combination of your talents, ability, knowledge, and intuition to bear in any one circumstance because of the continued distractions presented by other pressing priorities.

The best use of freedom is its disciplined exercise. The undisciplined use of freedom unfortunately can produce tumult in which, eventually, other people will determine your activities. That is the curious thing about discipline. You either exert it yourself or, by not exerting it, you eventually must submit to the control of others. Thus, you'll either discipline yourself or be disciplined by others.

As your productivity manifests itself, other well-meaning scientists will ask you to involve yourself in their projects. You must therefore strengthen your self-control so that it matches the increase in the number of opportunities for your professional involvement. First, develop the will to say no. Then develop the skill to say no.

Develop the ability to say no by recognizing that no damage is done to you when you exercise your right to refuse. As a faculty member, you're not obligated to take part in every project that comes your way. This fact is not weakened, but fortified when applied to junior faculty. As junior faculty, your time is even more precious during this critical formative period when you're developing new intellectual roots and grounding. Would you deliberately overwhelm or burden a colleague of yours, consciously reducing their ability to function effectively, single-mindedly degrading their performance? Don't do this to yourself.

> **The truth about discipline is that you either exert it yourself or, by not exerting it, you eventually must submit to the control of others. Thus, you'll either discipline yourself or be disciplined by others.**

If you choose to accept responsibility for a new task, do it with the recognition that there will be other new opportunities in which you can now play no role. When these other opportunities come to pass, choose to let them go by because of your prior commitment to the task at hand.

The greater the freedom, the greater the need for self-discipline. Retain control over how busy your days are. While it is important to push so that you get your work done for the day, don't frustrate yourself by insisting on being busy every moment of each day.

While it is important to ration your rest, choose to ration your work as well. Allow enough flexibility into your day so that you can read or have time to engage in discussions with colleagues. Permit yourself the opportunity to take advantage of small openings of opportunities that are of interest to you or to which you can make an important contribution without destabilizing your main efforts. You chose this career. It is perfectly all right to both work hard and to enjoy it.

Developing a Work Lifestyle

The key to this productive but flexible approach is equilibrium. Devote your energy to creating a balanced personal work lifestyle, developed and continually guided by character development. A personal work lifestyle is the atmosphere and surroundings in which you work. It consists of your own attitude, your workspace, your time, your calendar, and your colleagues. The development of this personal work lifestyle must appropriately weigh and integrate (1) your need to apply your professional talents to the fullest, (2) your need to be productive in science, and (3) your need to collaborate with fellow scientists, students, and administrators.

Your daily efforts should allow you to do all three of these. There is no doubt that it is a challenge to build this work lifestyle, but build it you must and build it to last. After all, you'll have a forty-year career or longer, and you want this structure to be one in which you're both comfortable and productive. Like any structure, it will need to be regularly inspected, repaired, and modified. However, this work lifestyle will pay a handsome dividend by providing the foundation for your personal and professional academic growth.

The core pillar of this work lifestyle is respect for yourself. Recognize with approval that you have a combination of talents that are of great value to the university. However, you cannot apply these talents effectively to any of your endeavors if you're harried by a schedule that is too busy or a workplace that is noisy and disheveled. Work for and insist on the environment that permits you the best opportunity to use the talents for which the university hired you. Then take advantage of this lifestyle by applying your strengths and talents to the didactic experience, research, and administrative problems at hand.

Discipline the use of your time so that you have the opportunity to critically think about important and central issues that come up at the university. While it remains important to avoid lackadaisical and tepid efforts that must be repeated to reach a competent standard, it is also important to shun overly zealous activity in one area that takes your time and attention from more important matters. Develop a light touch on the steering controls as you navigate yourself through your activities. Let your goal be diligent days, not exhausting ones.

Teaching and Character

A central role of faculty at a college or university is teaching. While there are libraries of materials on teaching styles, philosophies, and approaches, we will concern ourselves here with the development of an overall strategy for teaching and some tactical steps that you can take.

Developing a good teaching style begins with the recognition that students, their time, and their effort must be treated with respect and with dignity. As a new instructor, recognize that you cannot force students to learn. You instead must persuade them to teach themselves. You're not there to be their friend, nor are you there to intimidate or frighten them. Instead, you're there to persuade them to use the best of their talents to learn. This requires that you use all of your powers of persuasion to elicit from them the curiosity that they must bring to bear to master the course material. You'll have to apply the best combination of firmness and tact, of prestige and of empathy to persuade them to learn, absorb, and integrate the course material into their knowledge base.

This requires a consistent supply of energy from you as you deliver your lectures to these students. These students won't be challenged unless you challenge them, and providing this challenge requires a consistent and dependable supply of spirited power from you. Most importantly, this begins with approaching the next lecture with the best attitude.

Students are remarkably adept at determining whether you're tired or distracted. Their recognition that you wish to be somewhere else reduces the didactic experience for them because they believe that you hold them at a lesser-discounted value than they had hoped. Spending a few moments before class in emotional preparation for your lecture can provide handsome dividends for the class and often for you. It is sometimes useful to remind yourself that you're not just instructing students, but also future department heads, deans, judges, community leaders, and politicians. Those who will be charged with making controversial decisions in the future are gaining the foundation on which their decision will be made from your course of instruction.*

Keep in mind that you are only one resource from which your students can learn. Illuminate for them the other sources of information that are commonly available. One of them is a library. Many students treat a textbook like a spouse, that is, "to have and to hold until death do us part, forsaking all others." In fact, textbooks are useful, but imperfect tools. You'll no doubt have chosen a text for your class that is useful for your students, but invariably, students will react to that text with varying degrees of dissatisfaction. One student may find that the chapter on intracellular morphology may make the topic of molecular biology come alive, yet the following chapter on the instrumentation used by researchers may be terrible. Encourage students to broaden their reading for the course.

Spending part of the afternoon in the library will introduce the student to a number of texts, each with their own strengths and weaknesses. At the end of the course, the student will have a personal

---

* Getting into the habit and practice of giving clear, well-organized, and articulate presentations in class prepares you for giving clear and well-organized presentations before different and sometimes more hostile audiences.

library of material that taken together provides the wealth of material needs to cover the course.

Keep the students well oriented. Students appreciate well-outlined, well-designed courses. A clear and lucid layout of the course can make an important difference because students will not be compelled to take up valuable lecture time to discuss logistics of the course that could have been obviated with a clear statement of the course's prerequisites, lecture timing, content, and exam schedule. Remember that students commonly are challenged enough by the course material itself. They should not have to struggle to know what the material is that will be covered in the next lecture or when the assignments are due. Spending a considerable amount of time thinking about the design of the course before the course is offered can pay handsome dividends later in the semester.

Publishing and Character

We have discussed the need for publishing in peer-reviewed journals earlier in this chapter. In the university setting, you'll have ample opportunity to satisfy this requirement. Some useful guiding principles as you develop your strategy are as follows.

Recall that the purpose of your career is not to garner the maximum number of publications. It is instead to practice good science on a regular and consistent basis, building up your knowledge base and scientific sophistication, and integrating these with the development of your character. An important component of that character is intellectual honesty. This calls for an honest answer to the question, "Is this idea I have worthy of publication? Will it provide illumination of an important idea or concept?" If your critical appraisal suggests that the answer to this question is no, then suppress the paper. Discipline yourself to move forward into publication only those ideas, concepts, and analyses that you believe are worthy of dissemination to the scientific community. Doing otherwise is the equivalent of disrespect of your own good scientific judgment. When you make contributions, let them be contributions worth making. Your field of application should be better off with, and not cluttered, by your contributions.

You'll have the opportunity to make two types of contributions to the literature. The first will be based on collaborative effort. The second will be individual contributions that you can make. Typically, the collaborative manuscripts will come more easily. If you're competent, honest, articulate, and can easily admit your mistakes, you'll find that you have several opportunities to be involved in collaborative research efforts. These efforts will produce manuscripts fairly easily, and most commonly, these manuscripts will be authored by others more senior than you.

> **Your field should be better off with and not cluttered by your contributions.**

However, there are several disadvantages to collaborative manuscripts. First, they may not say what you as an individual would like for them to say. Collaborative manuscripts, by their very nature, are built on compromises that are made between the various authors, each with his or her own point of view. It is unlikely that you as a junior investigator will be able to have much of an influence on this complicated and subtle process. Your own point of view, regardless of how valuable, can get left "on the cutting room floor." Secondly, recall that by not taking a major role in writing a manuscript, your evaluators cannot gain a sense of your skill in written communication to other scientists.

For these reasons, give serious consideration to the idea of publishing a manuscript in your field for which you're the sole author. Taking such a tack holds several advantages. First, writing a sole-authored manuscript requires you to engage in a self-assessment in which you must identify the type of contribution that you wish to make to your field of application. Do you want to propose an idea or call for a specific class of analysis? Do you wish to expose a troubling style of research evaluation in your area of expertise and propose a replacement?

A most rewarding experience can be spending some time reviewing the literature, specifically examining the publications in your field. It can be illuminating to discover that while the literature itself is voluminous, the area of your specific and particular interest may not be well developed at all. This void can be filled by the contributory manuscripts that you wish to write. No one else has written them; they are for you to write.

Secondly, publishing your own manuscript requires you to develop your own writing skills. While it can be useful to observe the writing skills of others, work to develop your own by writing. This requires both time, practice, and patience. Progress comes steadily through practice. The best way to learn how to swim is to swim. The best way to learn how to write scientifically is to write scientifically.

If you have not published a manuscript on your own and are intimidated by the prospect, then start on a small project. Begin by authoring letters to the editors of some of the top journals in your field for publication. This activity serves a worthy purpose because editors benefit from the calibration of their decision-making process by the feedback they receive from their readers. Build your confidence and writing skills up slowly and steadily until you're willing to write a manuscript on which you're the sole author.

An advantage of working on a manuscript of your own is that you gain the valuable opportunity of taking complete control of the process. You and you alone are responsible for the text, tables, and figures. You and you alone decide on the submission date. You and you alone must respond to reviewers and negotiate with the editor.

Assuming and discharging these responsibilities requires growth and the development of new skills on your part. You'll gain valuable experience from the mistakes that you'll make. In all likelihood, not all of your single-authored manuscripts will be accepted. However, since the topic you have carefully selected to write about is worthy of dissemination, make your manuscript stronger with each round of reviews that it must complete. Finally, do not give up when you receive a rejection notice. Instead, expect rejection on your way to publication, much like a young ice skater expects to fall repeatedly as she perfects her skills.

The administrative and writing capabilities that are produced from these successful efforts will serve you well in the future. They will be particularly useful when you become first author of collaborative research projects. Your experience in selecting journals, your expertise in dealing with reviewers and editors, and your perseverance will ease your task of navigating the manuscript with its many authors toward publication.

Avoid ghost-authored manuscripts. A ghost-authored manuscript is an article that is written by individuals other than those listed on the masthead of the manuscript. These circumstances may occur when the true writers of the manuscript have a point of view that they would like to disseminate. However, since they do not have the same good and widely acknowledged reputation as others, they enter into an agreement with some of these other well-known scientists. These latter renowned scientists agree to have their names appear on the masthead of the manuscript even though they were not the researchers who authored the manuscript. They essentially are paid to have their names appear on manuscripts that they did not write. This fraudulent practice misleads reviewers, editors, and ultimately, the journal readership. It is intellectual hypocrisy, and its perpetrators' actions reduce the value of our profession.

Take Charge

Before you use your academic and intellectual freedom to drive your career, be sure that you first know where your career is going. As a junior faculty member, you'll have many opportunities from which you must choose. How specifically do you want to use them? One of the difficulties of being a junior faculty member is that you do not yet have the metric to decide which activities you should be involved in and which you should avoid. One of the most useful tools that you can develop to help you measure your available opportunities is to actively consider your long-term career goal.

How do you want to spend your time twenty years from now? Do you want to be a theorist, or will you involve yourself in practical applications? Do you want to go into administration? Do you see yourself eventually as a dean? Will you be a full professor with an active research program? Will you spend your days teaching students, or will you be out on the lecture circuit? Will you be involved in politics? Do you see yourself heavily involved in doing community service? Will you be teaching and guiding students? Character sabbaticals guide your development of answers to these questions.

Few families start a vacation trip without a vacation destination, wandering randomly and aimlessly, having their direction chosen by weather, traffic patterns, and spur-of-the-moment inclinations. Take some time now to plan your distant career destination. Pause to observe the senior faculty around you. Watch and ask them how they spend their days and time. Ask what they would do differently if they were starting out at the beginning of their career path, knowing what they know now. Use your mentor as a sounding board for ideas and concepts that you have for your future. Educate yourself about the possible obstacles that you'll face along that path so that if you choose that path, you'll be prepared to face them.

This is not a guaranteed path to success, but it allows you to move in a well-considered direction with your eyes well focused on your destination, being propelled by your own good talents and the intellectual and academic freedom that derives from being a faculty member.

23

Nova Progenies

It can be difficult to see how to put all this together. It's actually quite straightforward. Develop good instincts and trust them. If you are able to experience your character sabbaticals, you will be connected to the best parts of yourself. Sacrifice, vision, self-value, honesty, ethical point of view will be natural to you. Trust their leadings. When you are mistaken, admit it freely. It comes down to developing your character and then following your character's leads. In the new electronic era, our culture now recognizes that information devoid of insight is typically unhelpful and sometimes counterproductive.

It therefore comes as no surprise that society, in its attempts to integrate new data, prizes not just information, but information that is coupled with intuition and wisdom. We are progressing from asking the basic "What is it?" to the more illuminating "What does it mean?"

The contributions of science, commonly offered on a plate of increasingly technical and sophisticated experiments, are promising, controversial, and sometimes foreboding. Therefore, society will need the counsel of creative, insightful, disciplined, and mature scientists to guide it.

These are most likely capable researchers who both contribute to scientific progress and understand the implications of that progress. They must diligently work in science but must also possess solid principles and good judgment. They must be educable and of sound ethic. These are men and women who are both scientifically productive and of good character. It is my hope that this is the category of scientist to which you aspire.

The foundation for character growth begins with a solid, unshakable sense of your high value. There is only one person, and throughout the existence of mankind, there has been only one person with the unique combination of abilities, strengths, knowledge, curiosity, diligence, and intuition that you have. The instillation of these capabilities and insights within you first imbues and then empowers you with value separate and apart from external circumstances. External threats do not occur because you're valueless, but because you have value. Avoid falling into the traps of (1) believing that your value is determined by how you feel at any given time and (2) believing that your value is controlled by your performance. Generate within yourself a sense of worth that can neither be added to nor subtracted from. Don't be productive to generate self-value. Be productive because you have it.

Strengthening your development with character-building sabbaticals. These solitary honest reviews of who you are as a person, what you are doing, and why excite you to challenge your personal growth; to develop courage and compassion, ethics and honesty; and to prepare you for the price that you must pay for your principles.

If you can come to see that the reality of your elevated and constant value is greater than the reality of your day-to-day failures, then you can understand that no real damage can be inflicted on you if you fail in your diligent work efforts. This realization permits you to apply yourself with your full energy to your tasks, making bold efforts to learn, to teach, to grow, to think, to calculate, to postulate, to develop, and to mature. Scientific failures are temporary, and its defeats are transient while your high value retains its permanence. A solid sense of self-value permits you to be open to accepting the responsibility for leadership by protecting you from the damage produced by blame. Finally, it allows you to be secure enough to give the best of yourself to others, the hallmark of charitable prosperity.

As a junior scientist, first critically evaluate your strengths and weaknesses* with the view of building up your ability to deal with issues that have been a chronic problem for you. Work on and strive for self-transformation, converting your weaknesses into new positive energy. An affirmative sense of self-value permits you to stretch yourself to do what you don't feel like doing in order to become what you want to become.

---

* If you really cannot think of any weaknesses, ask your spouse or significant other for their perspective on the matter. They may have identified a few that might be helpful for you to know.

As a junior scientist, your environment is, and in all likelihood will continue to be, tumultuous. Many of your days will be unpredictable. You'll have to balance stability with adaptability, imagination with discipline, strength with compassion, intellectual exertions with restoration, professional priorities with personal ones. It is just as important to equilibrate these aspects of your life as it is to be productive.

Put another way, you must be productive in science while simultaneously developing balancing and coping skills in order to prosper in a chaotic environment. Investing all of your effort in productivity to the exclusion of character growth will fail you. Professional maturity requires that you develop both diligence and the ability to sensitively assess priorities calmly and unhurriedly during the daily cacophony of your job.

Do not ignore administration as you develop. The fact of the matter is that you'll spend an important component of your time doing administrative work. This time can be spent in two ways. You can spend it consumed by an attitude of resentment and frustration that will ultimately stunt the development of your project. Alternatively, you can be governed by a spirit and attitude of time generosity. As a diligent worker, you'll find the time you need once you have the attitude that you need. Don't just generate unhelpful experience; develop discerning expertise.

As a junior investigator, you only need your education and a solid, stable sense of self-worth to be a full participant in collaborative projects. With your sense of self-value that is separate and apart from your performance, you're free to participate and engage in the project, using all of your talents and abilities to support the group effort.

Don't shrink from, but instead, actively seek out opportunities to help other coworkers in the project. Be prepared to receive the unanticipated email or phone call that provides an opportunity for you to engage your talents and capabilities for the good of the project. When you make a mistake, apologize clearly and easily. Be willing to extend yourself, and even put yourself at risk, to provide support for a colleague.

The product of these efforts is a well-balanced, focused, and knowledgeable investigator who recognizes the importance of the project and communicates effectively with all the team members. Moreover, you'll also have the important combination of vigilance coupled with the willing attitude that permits you to shoulder not just your share of the work effort, but to help with the burdens of others as well.

Also, remember that it is up to you to inject some of the fun, satisfaction, and stimulation that pulled you to your science into your days. To that good end, insert a "productivity hour" into your day so that you might work quietly on a scientific matter of your choosing.

As a junior investigator, begin to deal at once with any presentation anxiety or stage fright that you have. If there is an "enemy" to be faced in giving a presentation, then that enemy is not the audience, but the fear of failure. If you're fearful, then confront that fear well before you get behind the podium. A strong sense of self-worth that is independent of your performance permits you to shine bright illumination on the dark shadow that is cast by fear, revealing that no lasting damage occurs to you if the presentation does not go as you hoped.

Choose to actively and affirmatively embed an ethic in your career. Ethics are more of an approach to life than a mere collection of rules. Your ethical behavior is the living expression of your core principles that govern your relationships with others. Just as you have self-worth on which you rely, your sense of the worth of others regardless of their opinions and actions governs your ethical treatment of them.

Take the opportunity to recognize that ethical researchers are not paragons. Ethical people make honest mistakes. However, what characterizes the ethical scientist is her response to those mistakes. When she recognizes that she missed an opportunity for ethical conduct, she apologizes, makes appropriate restitution, and having learned the right lesion from her error, she moves on. Ethical behavior is not perfect behavior. It is behavior that calibrates and self-corrects.

The skills that you develop as a junior scientist can naturally evolve into the foundation of good leadership. Begin to think of yourself as someone with leadership potential who only needs experience and tempering. Again, the key to successful leadership is security and a solid sense of self-valuation independent of your outward experiences.

Finally, you don't know what opportunities and dangers the coming day will bring. The small number of days that required me to make important and career-altering decisions began like every other day, providing no perceptible warning to me that I would be involved in an event or operation that would change my career. Be watchful for the unanticipated event that can alter your perspective and adjust your career for the better.

Finally, good luck. We are relying on you as a member of a nova progeny, a new generation of scientists.

Author's Background

Dr. Lemuel A. Moyé, MD, PhD, is a retired physician and biostatistician at the University of Texas School of Public Health. He earned his medical degree at Indiana University Medical School in 1978 and completed a PhD in community health sciences with concentration in biostatistics in 1987. He is a licensed physician in Texas and has actively practiced general medicine from 1979 to 1992. He is a diplomat of the National Board of Medical Examiners and is a retired professor of biostatistics at the University of Texas School of Public Health in Houston where he held a full-time faculty position. He has carried out cardiovascular research for thirty-four years and served as the coordinating center principal investigator for the NIH-Funded Cardiovascular Cell Therapy Research Network.

He has served in several clinical trials sponsored by both the US government and private industry. Dr. Moyé has served as statistician/epidemiologist for six years on both the Cardiovascular and Renal Drug Advisory Committee to the Food and Drug Administration and the Pharmacy Sciences Advisory Committee to the FDA. He has served on several Data and Safety Monitoring Boards that oversee the conduct of clinical trials and has participated in many reviews of grants that have been submitted by fellow scientists for federal funding.

Dr. Moyé has published over 220 manuscripts in the peer-reviewed literature that discuss the design, execution, and analysis of clinical research. He has published fourteen books. His latest is a 700-page treatise on Probability, Measure, and Public Health which is available for free at www.principal-evidence.com.

He has also authored *Face to Face with Katrina's Survivors: A First Responder's Tribute*, an account of his experiences as a first responder to the Hurricane Katrina catastrophe. This nonfiction book, published by Open Hand Publishing LLC, won the Skipping Stones, Foreward Magazine, Ed Hoffer, and Benjamin Franklin Book of the Year Award for Multiculturalism.

He has published three novels, is retired, and is living as a cancer survivor in Arizona with his wife Dixie.